VISIONS OF SCIENCE

VISIONS OF SCIENCE

Books and Readers at the Dawn of the Victorian Age

JAMES A. SECORD

The University of Chicago Press • Chicago

JAMES A. SECORD is professor in the Department of History and Philosophy of Science at the University of Cambridge, director of the Darwin Correspondence Project, and a fellow at Christ's College. He is the author of *Victorian Sensation: The Extraordinary Publication, Reception, and Secret Authorship of "Vestiges of the Natural History of Creation,"* also published by the University of Chicago Press.

The University of Chicago Press, Chicago 60637
The University of Chicago Press, Ltd., London

Printed in the United States of America

23 22 21 20 19 18 17 16 15 14 1 2 3 4 5

ISBN-13: 978-0-226-20328-7 (cloth)
ISBN-13: 978-0-226-20331-7 (e-book)
DOI: 10.7208/chicago/9780226203317.001.0001

LIBRARY OF CONGRESS CATALOGING-IN-PUBLICATION DATA
Secord, James A.
Visions of science : books and readers at the dawn of the Victorian age / James A. Secord.
pages ; cm
Originally published by Oxford University Press, 2014.
Includes bibliographical references and index.
ISBN 978-0-226-20328-7 (cloth : alk. paper) — ISBN 978-0-226-20331-7 (e-book) 1. Science—Great Britain—Historiography. 2. Science—Great Britain—History—19th century. I. Title.
Q127.G4S44 2014
507.2'2—dc23

2014010721

⊗ This paper meets the requirements of ANSI/NISO Z39.48-1992 (Permanence of Paper).

For Anne

CONTENTS

PREFACE

When we imagine the future, we think about science. In this book, I suggest that this distinctive way of projecting what the novelist H. G. Wells later called 'the shape of things to come' first became widespread during the upheavals of the early industrial era in Britain, and especially in the turbulent decade around 1830. During these years, science was changing from a relatively esoteric pursuit into one known to have profound consequences for the everyday life of all men and women. A key feature of this transformation was the flourishing of books reflecting on the practices and prospects of science. These books range from polemics about national decline to overviews of the unity of knowledge. Among a host of potential titles I have selected seven of the most important. These works have long been celebrated within particular disciplines, but the significance of the utopian moment that produced them has been forgotten. They were not just accessible versions of specialist research, what we pigeonhole today as 'popular science'. Rather, these books offered comprehensive perspectives on science and its meaning for human life, at a moment when the first industrial nation appeared on the brink of revolution.

At this extraordinary juncture, the country confronted the prospect of a greatly expanded reading public. Reading had the potential to define the character of the nation. Who should read what, and towards what end, was fiercely debated. What would hold a reformed political order together? How were people to define their identity in an era that looked likely to be dominated by global commerce, a class-based society, and life in cities? These questions would come to be asked throughout the world wherever traditional societies faced the challenges of the new order of industry and empire. For many readers, the answers were to be found in books about science. The heroes of my story are these books and their readers.

As a book about books, *Visions of Science* is first and foremost indebted to libraries and librarians. An invitation from Anne Jarvis and Jill Whitelock of Cambridge University Library to deliver the Sandars Lectures in Bibliography for 2013 gave shape to the project. The opportunity offered by David McKitterick to contribute to *The Cambridge History of the Book in Britain* (2009) made it possible to develop some of my ideas. I wish especially to thank Anna Jones and Tim Eggington of the Whipple Library, as well as Dawn Moutrey and Steve Kruse for help with images. The staff of the University Library in Cambridge has been uniformly helpful over many years, particularly those in Special Collections, as have librarians at the National Library of Scotland, the British Library, the Bodleian Library, and the Library of the Royal Society of London.

I am grateful to colleagues, students, and staff in the Department of History and Philosophy of Science at Cambridge both for intellectual stimulus and generous provision of leave. I am also deeply indebted to my colleagues on the Darwin Correspondence Project, particularly Alison Pearn, for making it possible for me to finish this book. This work has been enriched by my participation in the Past versus Present project of the Cambridge Victorian Studies Project and the Generation to Reproduction Project, funded respectively by the Leverhulme Trust and the Wellcome Trust. I am also grateful to Gareth Stedman Jones and Inga Huld Markan for an invitation to contribute to the seminar series on History and Economics at Cambridge, and to Eileen Gillooly and Jonah Cardillo for the opportunity to speak to the Society of Fellows at Columbia University.

My knowledge of the apocalyptic image of the 'March of Intellect' comes from Alison Winter, who long ago convinced me to think about this period in a fresh light. Jan Golinski, Ralph O'Connor, and Clare Pettitt provided critical readings of chapters. Over many years I have benefited from conversations with Simon Schaffer, Jack Morrell, Nick Hopwood, Adrian Desmond, Aileen Fyfe, Bernard Lightman, Adelene Buckland, Katharine Anderson, Gillian Beer, Duncan Bell, Jane Munro, Martin Rudwick, Mary Beard, Evelleen Richards, Boyd Hilton, Peter Mandler, Simon Goldhill, Paul White, John van Wyhe, and Richard Yeo. A special thanks goes to Jon Topham, whose unparalleled knowledge of science and medical publishing in this period has transformed my understanding. I am grateful for the bracing criticisms and helpful suggestions of the academic referees for Oxford University Press, and to Karen Darling and the University of Chicago Press for their interest in the book. Working with my editor, Latha Menon, and the team at Oxford has been a pleasure throughout.

Through her critical insights and constant encouragement, Anne has shaped every aspect of this book: she is my best reader.

LIST OF FIGURES

LIST OF PLATES

Introduction

> We are assembled here in performance of our part of a process... on which it is no exaggeration to say, that the future destinies of this empire will very mainly depend.
>
> —John Herschel, 'An Address to the Subscribers of the Windsor and Eton Public Reading Room', 1833

On the evening of 29 January 1833, the astronomer John Herschel addressed a meeting in the Christopher Inn on the high street of Eton, a village on the outskirts of London. The only son of the planetary discoverer William Herschel, he was already a celebrated man of science in his own right. Herschel was speaking as President of the Windsor and Eton Public Library and Reading Room, which had been recently founded to spread the cost of books to the local community. Those at the meeting had just agreed to offer a reduced subscription for 'industrious' readers, but Herschel urged his listeners to go further by providing a free selection of books to even their poorest neighbours. The universal spread of reading, Herschel believed, provided the best hope for raising 'the standard of moral and intellectual culture in the mass of the people', which would in turn lay the foundations for progress, civil liberty, and national unity.[1]

Fewer than a hundred heard Herschel speak in the old coaching inn, but his words were soon published and quoted across the world. This was a utopian moment, for nothing less than the future of civilization appeared at stake. During the previous decade Britain had come as close to revolution as it had at any point since the seventeenth century, with mass demonstrations over the right to vote, the removal of restrictions on religious minorities, and the replacement of human labour by machines. The Reform Act, to which Herschel alluded in his speech, had passed through Parliament in June 1832, clearing away the worst abuses of the old political system. But the reform debates had revealed a nation on the edge of an abyss, with riots, arson, and machine-breaking. Preachers foretold the approaching end of the world from pulpits, and philosophers characterized the 'spirit of the age' in weekly magazines. Herschel's quiet and considered remarks were part of this prophetic literature, as he looked forward to a future when the lessons of great literature—rather than religious or political dogma—would guide public debate. Readers could start with Miguel de Cervantes' *Don Quixote* or Samuel Richardson's *Pamela,* and, thus prepared, could move on to the 'higher' realms of poetry, history, and philosophy.

The sciences, increasingly seen as a touchstone of rational endeavour, had a uniquely important role to play in this new order. This was brought out at a subsequent meeting of the Windsor and Eton subscribers by the publisher Charles Knight, who from local origins as a printer in Windsor had become the nation's leading purveyor of inexpensive non-fiction works. As Knight stressed, the spread of this kind of 'useful knowledge' would provide readers of all classes with a standard for measuring the words of demagogues, gossips, and politicians.[2] During the late 1820s and early 1830s, men like Herschel and Knight were at the forefront of a movement that stressed the reading of science as a remedy for the country's social, political, and religious malaise. Conversation about science pervaded elite society, new institutions for practising and disseminating knowledge were founded, and hundreds of thousands of men and women paid a penny a week to gain access to chemistry, astronomy, and other sciences. The reading of science appeared to possess the power to change not only the way people read, but their innermost thoughts and actions.

At the height of these transformations, a series of remarkable books reflected on the nature of the sciences and their meaning. Each chapter of the present book explores the making and reception of one of these works as a focus for public controversy. Growing out of the hopes for the power of knowledge, each of them projected a vision of the future. Some, such as Herschel's *Preliminary Discourse on the Study of Natural Philosophy,* outlined how the sciences could be used to define good character and appropriate ways of behaviour. Others, such as George Combe's *Constitution of Man* and Charles Lyell's *Principles of Geology,* established foundations for controversial subjects—the human mind and the long history of the earth. Still others, notably Mary Somerville's *Connexion of the Physical Sciences,* developed new literary forms to show that the sciences could be unified and law-like without leading to materialism. Humphry Davy, a leading chemist of the previous generation, used his last work, *Consolations in Travel,* to contemplate questions of permanence, immortality, and change across vast reaches of time and space. *Reflections on the Decline of Science,* by the mathematician and aspiring politician Charles Babbage, offered a withering critique of contemporary scientific institutions. And finally, an anonymous magazine serial by a lapsed mathematical teacher, Thomas Carlyle, suggested that the route to understanding our place in nature would be discovered not in astronomy or chemistry, but through a philosophy of clothes. But like the books it parodied, 'Sartor Resartus' (the 'tailor retailored') imagined new foundations for a scientific understanding of nature.

These seven books are by no means the only works of scientific reflection to have been published in these pivotal years, but they have long been acknowledged to be of special importance.[3] They deserve to be widely read, and understood together as the outcome of a unique moment of change. For the first time in history, the consequences of a scientific understanding were being developed not just for wealthy and privileged elites, but for a wide range of men and women, particularly in the middle classes and upper reaches of the working population. In making knowledge available to new readers, these books were a response to the possibilities of innovative technologies of production and distribution. As the British Empire's network of trade, influence, and military conquest expanded during the coming century, the process that had begun in villages like Windsor and Eton was repeated from Sydney and Shanghai to Cairo and Calcutta. Nearly a year after speaking in the inn at Eton, Herschel arrived at the Cape of Good Hope to survey the skies of the southern hemisphere, where he initiated a programme of public education and astronomical lectures.[4] Throughout the world, the arrival of the modern age was signalled through the establishment of libraries, printing presses, and ambitions for mass literacy in science.

The Power of Science

It was an age of signs and wonders, in which, as one commentator noted, 'events were visibly marching forward to that great visible era of doom and triumph'.[5] Britain was a Christian country, steeped in the language and thought of the Bible, and the arc of history was generally thought to end in the millennium, when Christ would return to rule in glory prior to the Last Judgment. Crowds flocked to see the visionary paintings of John Martin depicting the 'Destruction of Nineveh' and 'The Deluge'. At the same time, there was a sense of limitless possibility through projections of the future economy based on machines, and especially the application of steam to transport and manufactures. The opening in September 1830 of the first railway linking Manchester and Liverpool was widely hailed as marking a new era. The steam-powered future of the year 2000 was imagined in a print

(Plate 1) entitled 'The Century of Invention', with balloons, perpetual motion machines, carriages, and even self-propelled houses that travelled on the rails. 'Tomorrow evening...' the placard of a figure in the foreground announced, 'a cast iron parson will preach by steam at Fudge Chapel'.[6]

Utopian hopes for the new age were embodied, above all, in science. From the late eighteenth century onwards, the institutions, practices, and ideas of science had been undergoing a transformation. The change is often compared to the familiar 'scientific revolution' of the seventeenth century, associated with the names of Copernicus and Newton, but is more appropriately seen as the invention of modern science itself. Old institutions and methods of understanding were recast and new ones developed that we now see as fundamental. Science (based on the Latin *scientia*) had previously included all theoretically grounded knowledge—including grammar, rhetoric, and theology; now it was increasingly used to include only the study of nature. New institutions were organized for conducting investigations, with specialized facilities including batteries, furnaces, and other instruments for breaking things down and combining them in unexpected ways. Pre-eminent among them was the laboratory, which had been used in chemical studies but now became a place to explore phenomena from magnetism and light to digestion and nervous action. Museums of natural history, which since the Renaissance had been places for display and wonder, were becoming sites for contrast and comparison of the inner workings of plants and animals. In the study of magnetism, light, and other physical phenomena, the transformation involved the invention of 'physics' with a new relation between mathematics and experiment. Electricity, which since the Greeks was known only through sparks and lightning, was now available from batteries as a current in a wire. In the investigation of the earth, a range of activities from mineral collecting to Scriptural exegesis were forged into the new science of geology, with a focus on the mapping of strata and their history. Animals were viewed from a functional perspective, with all of their parts contributing to a working organism. Behind all these new approaches was a focus on analysis, in which the natural world was taken apart like a machine.[7]

The role for the enquirer into nature was also being transformed, from older images of scholarship and learning to the new ideal of the heroic discoverer, engaged single-mindedly in the investigation of nature. Among the best known was John Herschel's father, William Herschel, whose assiduous scanning of the heavens had led in 1781 to the discovery of the first planet since antiquity, Uranus. Another example, as we shall see, was the chemist Humphry Davy, whose use of the electrical battery led to the discovery of chemical elements and the new field of electrochemistry. Yet another discoverer was John Herschel, with his pioneering astronomical work at the Cape. With this novel role for the natural philosopher came new images of scientific authorship and new forms of scientific prose, from articles in specialist scientific journals to books reflecting on the consequences of science for everyday life.

The new practices that developed throughout Europe in this period provided science with an unprecedented potential for authority over interpretations of nature, an authority which many saw as underpinning the realms of politics and religion (Fig. 1). This form of control mattered because some of the new scientific findings were not only exciting, but also potentially threatening to established institutions and traditional ways of thought. In the first two decades of the nineteenth century, the public image of the sciences continued to be dominated by longstanding controversies in philosophy, politics, and theology: understanding of the earth required interpreting the Creation and Flood stories in Genesis; contemplating the heavens involved questions about the formation of the universe and the possibility of life on other planets; and investigations into matter entailed questions of life, mind, and spirit. These were all debates

Fig. 1. This woodcut vignette, part of a conservative reply to the radical press, suggested that the nation's true stability lies in constitutional principles, the laws of England, and the Bible, all of which rest on foundations in nature provided by the Newtonian system of the universe. Across the Channel, these supports are lacking and the French nation crumbles. 'Order is Heaven's First Law', from *The Real Constitutional House that Jack Built* (London: J. Asperne, 1819), vignette from title page.

conducted with reference to wider religious and political issues, even when focused on observations and experiments. From this perspective, discoveries in science could be used either to support or challenge the established order. The indication of vast ages in the past uncovered by geology seemed to contradict the literal truth of Scripture, and although there were readily available strategies to remove this threat, such findings were widely used in attacks on Christianity. The study of chemistry, it was feared, could easily lead to materialism—the doctrine that everything in the universe, including mental and spiritual phenomena, was the result of matter in motion. In the anatomy schools of Edinburgh and London, young medical students flocked to hear life described as nothing more than a special property of the organization of matter. This was the kind of science that had fascinated the young Mary Shelley, who relied on it in *Frankenstein*.[8]

As the case of *Frankenstein* suggests, concerns about the misuse of new forms of science were very real. Shelley was fully aware of the latest anatomical work, which, like many of the most exciting and dangerous innovations, came from Paris. Military conflict with the French may have ended with Napoleon's defeat at Waterloo in 1815, but the war of ideas grew even more intense. Paris was the scientific capital of the world in the 1820s. Its museums had the best specimens, its instrument makers were unrivalled, and its lecture halls were filled with students from across Europe, drawn by accounts of the latest discoveries. With these resources at their disposal, savants wrote comprehensive works on their subjects, bringing together findings from around the world and from decades of specialist research. The greatest work on fossil shells, by the naturalist Jean-Baptiste Lamarck, was prefaced by a scientific argument for the origin of new species though an evolutionary process of transmutation. Most notoriously, Pierre-Simon Laplace's magnificent five-volume *Mécanique céleste* never mentioned God—for Laplace believed that a mathematically complete system of planetary motion did not require divine interference.

The implications of such findings were greatly feared in Britain, where the shock of the French Revolution, with its guillotines and godless libertarianism, remained a potent reminder of what could happen when the old truths about nature were abandoned. The fears were made real by agitators like the tinsmith Richard Carlile, who employed the latest scientific findings in campaigns to destroy the theological bonds used to suppress ordinary working people. In inexpensive, seditious, and often illegal publications, geology, astronomy, and other sciences could be brought to the service of materialism and atheism. In such circumstances, was it safe to encourage ordinary working people to read science at all?

The question was particularly pressing because of the widespread participation in science in Britain, with practitioners ranging from genteel women fossil collectors to artisans who built their own telescopes and globes. The full-time pursuit of science was usually limited to men with independent incomes or by those with wealthy patrons. Apart from the few curatorships or university chairs, science was not a paid career in Britain, as it was in France, but a vocation, in which devotees gathered together to discuss issues of common intellectual interest. The most important collections of books and specimens belonged not to the state, but to wealthy gentlemen. Results tended to be reported piecemeal, and informal networks of conversation, letter-writing, public lecturing, and discussion remained at least as important as the publication of discoveries for priority. By the 1820s, however, these older forms of organization were being challenged by groups such as the Geological Society of London (founded in 1807) and the Royal Astronomical Society (founded in 1820), in both of which a small number of gentlemen with specialist expertise coordinated a wide range of observers. The new specialists were concerned to establish boundaries around their enquiries, so as to make their subjects free from accusations of speculation and infidelity.

The traditional way of demonstrating the compatibility of science with the established order was through the evidences of divine design, to show that nature could be read as God's second book. But how to do this remained controversial. Most orthodox Protestants tended to look for an overall plan in nature, a universe guided by divine law; yet too great a stress on laws of nature could seem to move God further away from the Creation, or could even justify atheism. More ardent evangelicals searched for evidence of divine intervention or parallels between nature and Scripture. The greatest challenge was presented by Christ's sacrifice on the cross and the fate of the individual soul, about which natural theology had little to say. The most widely circulated work in the tradition, William Paley's *Natural Theology* of 1802, presented a God that for many was too much like an architect of matter, a designer distant from his creation. Natural theology would remain important in Britain through much of the nineteenth century, and one of the most successful series of reflective works in the sciences, the Bridgewater Treatises, aimed explicitly to use the latest science as evidence of the wisdom and goodness of God in the Creation.[9] Such works, however, gained their power only if the findings they reported had been made independently of the theological conclusions to be drawn from them; hence they enhanced the significance of geology, stellar astronomy, comparative anatomy, and the other scientific disciplines on which they were based.

Portents of the Future

These profound hopes and fears about science emerged out of an unparalleled concern with the future, as portents of disaster could be read every day in the newspapers. In December 1825, the banks crashed in the first great financial catastrophe of the modern era.[10] In 1829, faced by the prospect of open insurrection in Ireland, Parliament removed legal restrictions on Catholics; perhaps this was a sign of increasing liberality, but it might give succour to potential followers of the papal Antichrist. In 1831 a congregation at Port Glasgow in Scotland witnessed the 'extraordinary manifestation' of speaking in tongues. Was this divine inspiration, a pious fraud, or a symptom of collective madness? Later that year 'King Cholera' arrived in London, killing thousands in its triumphal passage. These cataclysmic events were considered providential judgments, seen by some as punishing the sins of individuals, by others as the catastrophic failure of the entire nation. Above all, there was a call for reform of outdated institutions and practices of 'Old Corruption': the maze of sinecures, nepotism, places, and pensions at the heart of Church and State. Radical critics pointed to the need for change, from the parish churches to the prisons, from the courts to the learned institutions.

No institution was thought more in need of reform than Parliament itself: notoriously, uninhabited ruins were represented in Westminster, while fast-growing industrial towns such as Manchester, Birmingham, and Leeds had no representation at all. In a population of some 14,000,000 only 400,000 could vote, and how the franchise might be expanded—or if it should be changed at all—was violently debated. In general, the Tories opposed reform, the 'Ultras' completely so, while the Whigs supported it and the Radicals hoped to go further than the changes being proposed. Popular pressure to change the composition of Parliament was intense, with riots in cities throughout the country after the House of Lords rejected the reform bill passed by a recently elected House of Commons. When a revised bill finally did pass in 1832, it allowed for male suffrage with a ten-pound property qualification. The new industrial towns were enfranchised, the number of voters increased by about 60 per cent, and the most corrupt features of the old system of representation were removed. The most serious concerns of the middle classes had been met, at least temporarily. However, the basic principle that voting should continue to be based on interests in property, rather than rights as a citizen, was deeply conservative. Those with insufficient property, or those (such as women and most working people) considered to be under the sway of others, were deemed not to have a sufficient stake in maintaining social order. For radicals hoping for a wider—not just an increased—suffrage, it was a bitter defeat.

Even before the limited nature of reform was clear, the country was in danger of being torn apart. What could be done? To many observers, the rational understanding provided by science offered a way forward. The key figure in this movement was Henry Brougham, a

Fig. 2. The book-filled head of Henry Brougham, who was best known as an author in connection with the Library of Useful Knowledge and the *Edinburgh Review*. 'A Box of Useful Knowledge' (Lambeth: Thompson, *c.* 1832).

Whig politician instrumental in reform, who had written extensively on mathematics and the sciences (Fig. 2). In the early 1820s Brougham became something of a hero to radicals when he defended Queen Caroline against the future George IV at a notorious trial, which demonstrated the political power of the printed word in reaching all

Fig. 3. A transparency of this image of Liberty and the press was displayed during the illuminations in London celebrating victory in the trial of Queen Caroline. Liberty holds a medallion with the Queen's image in one hand, and a pole with a French liberty cap in the other. Behind her is a wooden-frame printing press. Their combined light drives away the forces of aristocratic and Church privilege. The words 'Triumph of the Press' and 'Knowledge is Power' were displayed as mottos in the same display. George Cruikshank, 'The Transparency', from William Hone, *The Political Showman—At Home! Exhibiting his Cabinet of Curiosities and Creatures—All Alive!* (London: William Hone, 1821).

classes (Fig. 3). Brougham was well aware that working men, from Glasgow to London, had been forming independent clubs and mechanics' institutes to reap the benefits of self-education. Many working-class readers looked to intellectual improvement as a route to political independence. Practical journals edited by practical men, often with a radical political slant—such as the *Chemist* and the *Mechanics' Magazine*—were providing information about the latest inventions and discoveries. The engraver and poet William Blake, with his hatred of 'the keen polar atmosphere of modern science', may have condemned the first issue of the *Mechanics' Magazine* as an enemy to art, but most of his fellow artisans did not agree.[11] In 1825 Brougham trumpeted forth his views in a best-selling tract, *Practical Observations upon the Education of the People,* which recognized the developing culture of print among working people and aimed to turn it towards moderate reforming ends among the population as a whole, including the middle and upper classes.

Although *Practical Observations* specifically targeted literate workers and their masters, the aims of what became known as 'the march of intellect' were much broader. For one thing, only about half of the men and women of England could sign their name on a marriage register, so the proportion of the working classes involved in any kind of self-improvement through reading was inevitably restricted.[12] Most importantly, it is rarely recognized that the debate about learning initiated by Brougham was not aimed only at the labouring classes, but at everyone, and involved all levels of education from studies at university to the acquisition of basic literacy.[13] The goal was nothing less than the complete reformation of society through knowledge, which, as Brougham said in a speech at the University of Glasgow later that spring, 'must put to sudden flight the evil spirits of tyranny and persecution, which haunted the long night now gone down the sky.'[14] 'Knowledge', as the slogan revived from the seventeenth-century essayist Francis Bacon put it, 'is power', and that power could be achieved by self-education in science. A reading public would be the basis for a reformed British polity.

The Mechanisms of Intellect

Behind these millenarian hopes in reading was a faith that machine production would transform the availability of knowledge. As Knight never tired of pointing out, his contemporaries were living through one of the great revolutions in communication, with technical innovations replacing practices that had changed little since the introduction of printing in the fifteenth century. The steam-powered printing

Fig. 4. An Applegarth and Cowper printing machine, powered by steam and engaged in printing the *Penny Magazine* of the Society for the Diffusion of Useful Knowledge. A wood engraving from 'The Commercial History of a Penny Magazine', *Penny Magazine* (31 December 1833), 509.

press vastly increased the number of impressions that could be produced every hour, making it possible to publish books and periodicals at much cheaper prices (Fig. 4; Plate 2). Paper, which for centuries had been made by hand a sheet at a time, could now be machine-made in rolls. Similarly, the use of stereotype plates in plaster or laminated paper ensured that additional printings of a work could be produced without the need for resetting type. These new technologies, however, required capital on a scale that only middle-class entrepreneurs like Knight could afford, which removed at least some of the dangers associated with distributing the sciences widely. Rabble-rousing demagogues like Carlile and William Hone, still relying on production by hand, could be priced out the market.[15]

The new entrepreneurs of knowledge aimed to provide reliable and standardized knowledge, which they presented as rising above the usual sectarian and ideological divides. Dry concepts and isolated facts were to be brought to life by their relations to one another, through the process of reading and its integration into everyday life. Brougham drew on his persona as a famously busy person to make the point that anyone could spare a few hours for reading. His *Practical Observations* asked labourers to save three pence from their weekly wages to buy Benjamin Franklin's classic account of his rise from a humble printer's apprentice to electrical discoverer and man of affairs. The Franklin autobiography was the product of one of the pioneers in the field of cheap publishing, John Limbird of London, best known for *The Mirror of Literature, Amusement, and Instruction,* a pioneering two-penny weekly. If a working man would just look at the first page, Brougham declared, 'I am quite sure he will read the rest; I am almost quite sure he will resolve to spend his spare time and money, in gaining those kinds of knowledge which from a printer's boy made that great man the first philosopher, and one of the first statesmen of his age.'[16]

To provide such knowledge on a large scale, Brougham founded the Society for the Diffusion of Useful Knowledge in 1826, a year after his original pamphlet. Although mocked as a Whig folly, the Society quickly became a cross-party coalition of moderates, a model of how the reformed political order might function, uniting metropolis and province, rich and poor. The Society's annual report for 1827 confidently predicted that they would 'leave nothing undone, until knowledge has become as plentiful and as universally diffused as the air we breathe'.[17] Rather than achieving cheapness by reissuing old books, it commissioned new ones and published them in weekly parts at the lowest possible cost, so that the latest knowledge would be available to readers of all classes, from dispossessed handloom weavers to the gentry and aristocracy. Topics ranged from brewing and botany to hydrostatics and hydraulics, with a later series including lighter 'entertaining' topics such as insect habits and Egyptian antiquities. Its most successful venture was the weekly illustrated *Penny Magazine*, which at its peak achieved a circulation of over two hundred thousand copies. The closely packed publications, filled with tiny print, which resulted from these initiatives were criticized as incendiary by conservatives, as useless by radicals, and as ridiculously utopian in many novels and newspapers.

These innovations in publishing made readers acutely aware of qualities of binding, paper, and printing. Size mattered, with small octavos (which involved folding the printed sheet three times) tending to be less costly and prestigious than more generous folios, quartos, and large octavos. Machine printing did not allow for the same fineness of production that could be achieved by the use of the traditional hand press, and the need to cut costs often resulted in small type in double columns on crowded pages with narrow margins. The very cheapest publications appeared in wrappers or simple folded sheets, and were usually published in parts to spread the cost of purchase. Mid-priced books might be ready-bound in a cloth casing, an innovation that made it possible for purchasers to avoid the extra expense of rebinding. Books could also be rebound in leather, an alternative that implied status, permanence, and individual taste. Caricatures such as 'The Pursuit of Knowledge under Difficulties' (Fig. 5) showed what happened when different formats of books and classes of readers collided, as a cabinet-maker absorbed in a political newspaper crashes into a gentlemen with an elegant pocket volume in hand. Pursing a similar theme, the caricaturist Robert Seymour depicted 'The March of Intellect', parodying the way in which readers of the wrong kind were being encouraged to read the wrong works in the wrong places in the wrong way (Plate 3). A busy street in St Giles, notoriously the worst slum in London, is blocked by pursuers of science and art, with the walls plastered

Fig. 5. A gentleman reading an elegant pocket volume is hit by a cabinet-maker reading a political sheet. The title of this lithograph, 'Pursuit of Knowledge under Difficulties', had become a catchphrase after publication of a book with that title as part of the Society for the Diffusion of Useful Knowledge's Library of Entertaining Knowledge (London: Thomas M'lean, 1834).

with advertisements for useful knowledge publications. The very idea of a 'St Giles's Reading Society', as shown on the cover of one volume, is intended to raise a laugh. The public was amused, alarmed, and fascinated by the prospect of universal literacy, an issue at the leading edge of technological change.

Shortly after Brougham's campaign began, the publisher Archibald Constable of Edinburgh outlined his plan for inexpensive books in a drunken conversation at Sir Walter Scott's baronial mansion in Scotland: 'a three shilling or half-crown volume every month which must and shall sell, not by thousands or tens of thousands, but by hundreds of thousands—ay, by millions'. Scott predicted that Constable would become 'the grand Napoleon of the realms of *print*'.[18] In fact, Constable went bankrupt in the financial crash of the mid-1820s, but other publishers saw an opportunity to reach new markets with small cheaply printed books. The five-shilling small octavo format used in Scott's hugely popular collected works became a standard for publishing not only in history and biography, but for science as well. Constable's vision, revived as a miscellany, included Gilbert White's *Natural History of Selborne* in two paper-covered parts at a budget price, introducing the work to readers who could not afford the earlier, expensive editions on fine paper with large type.[19] Following Constable's example, John Murray published a 'Family Library' in 47 volumes and Longman issued a 133-volume *Cyclopaedia*. Such books were cheap by the standards of the larger formats that had previously dominated the trade, but the price still put them out of the reach of all but the wealthiest 20 per cent of the population. They were, however, staples of subscription libraries such as the one addressed by Herschel. Over the longer term, the most successful entrepreneurs in the field of really inexpensive publishing were William and Robert Chambers of Edinburgh, whose state-of-the-art printing establishment led the campaign for the enlightenment of the middle and working classes.[20]

These transformations in publishing were accompanied by a remarkable fluidity in the boundaries between different types of writing. The years around 1830 were traditionally depicted in literary history as a hiatus between the Romantics and the Victorians—between the death of Lord Byron in 1824 and the serial publication of Charles Dickens's *Pickwick Papers* in 1836—but are increasingly understood as an extraordinarily fruitful period for experiments in prose. Verse and traditional fiction were seen by contemporaries as in decline, with publishers and readers looking to the dialogue, the essay, and the treatise as indicators of what William Hazlitt termed the 'spirit of the age'. This is the era of the great historical essays by Thomas Babington Macaulay and Thomas Carlyle in the *Edinburgh Review*, the satiric novels of Thomas Love Peacock, and Mary Shelley's futuristic fantasy *The Last Man*. Robert Southey's *Sir Thomas More: Or, Colloquies on the Progress and Prospects of Society* of 1829 provided two volumes of dialogues, illustrated with engravings of the landscapes of the Lake District, in which a Tudor humanist time-travels to contemporary England to discuss steam engines and Catholic emancipation. Edward Bulwer-Lytton's *The Last Days of Pompeii*, published in 1834, was a thrilling adventure story that doubled as Christian apologetic and an archaeological guide to the remains of the ancient city.[21]

Among these literary experiments, some of the most significant and successful were books reflecting on the wider meanings and possibilities for science. These works tamed fears about the implications of new discoveries and shaped scientific practices into forms that remain with us today. They provided flash points for discussion about the future role of knowledge in society, generating a lively outpouring of reviews, extracts, letters, lectures, sermons, essays, and pamphlets, which in turn reached far larger readerships. These books created worlds; they traced the contours of disciplines, outlined epics of creation, and projected the reader into the universe of the infinitely small or large. They explored new forms of writing, new tools of enquiry, new relations between matter and spirit, and new roles for the individual. Taken together, they established a canon that lasted through the Victorian era and are still regarded as founding texts for subjects from psychology to physics.

Feeding the Little People

In February 1828, having just learned that the Duke of Wellington, reactionary hero of the Battle of Waterloo, was to be the next prime minister, Brougham declared in Parliament that a greater force was in the land. Books were more powerful than bayonets; 'the schoolmaster is abroad'. The newspapers mocked Brougham's epic ambitions in language drawn from the Book of Revelations: 'He is the *Solomon* of science—the master of mechanical systems. . . . The day is at hand when he shall stand forth the Great Captain of the Age, and at the head of his legions begin the march of intellect.' One article nervously joked about what would happen if the plan for universal education in science failed:

> If they won't read my boys, then make them bleed my boys:
> Should they turn tail, then their tails you must tickle.[22]

Opposed by conservatives, despised by radicals, ridiculed even by some reformers, the movement to unite the nation through the reading of science transfixed public attention. In the wake of Brougham's speech, another large caricature by Robert Seymour in the series entitled 'The March of Intellect' (Plate 4) went on sale in London's print shops. 'I saw a Vision', the caption begins, of a 'giant form'. The monster has eyes of gas-lights and the body of a steam printing press. Its head is made of books, its crown is the dome of the newly built London University, and in its hands is a huge broom, a pun on 'Henry Brougham'. Swept away are quack doctors, lawyers, corrupt clergymen, and a huge mass of useless printed material.

'I come, I come!' cries the great machine of intellect, ushering in the universal availability of information. As with transformations in electronic communication today, new technologies sweep the old into dust.

Yet examined more closely, the relations between humans and machines in this image are ambiguous and intertwined. Machines were not something separate, but were extensions and integrations of human labour. The setting of type remained among the most skilled of all manual trades; presses run by steam demanded constant monitoring; paper might be made by machine, but pages needed to be folded, sewn, and bound by hand. Even the title of the print 'The March of Intellect' is based on an anthropomorphizing analogy, in which literal armies of soldiers and gun carriages are transformed into figurative ones of printing presses and books. Brougham seems a helpless instrument of the monster he has helped create; but the creature's head is composed of volumes of the kind that Brougham himself had written or promoted. The print itself is made on paper hand-made from rags; it is coloured by hand, and it is an etching, a form of printing that only allows a limited number of copies. The image would have been viewed largely by the wealthy, in books of prints intended to spur conversation. The most fearful aspect of the print is that the machine is moving with intent: matter is vibrantly and dangerously alive, and pouring forth from it are the 'small Books that fed the little people of the Earth'.[23]

This book is the story of these books. It will show that works reflecting on the wider meaning of science, like all literature, can be understood through close reading and an understanding of their physical qualities as books, in light of the experiences of those who bought, borrowed, and discussed them, from women at metropolitan soirées to schoolchildren rejecting geological views of earth history. It will uncover the role of printers and publishers, from factories pouring out cheap compendia to fashionable premises in London's

West End. Although the books exploited innovative formats and often targeted broad readerships, they were not popularizations, digests of existing research in an accessible form.[24] In various ways, they can be defined as reflective treatises, considering the meanings of science and its place in modern life. They looked to the future, coordinating and connecting the sciences, forging a knowledge that would be appropriate for the new age. Their aim was often philosophical—to determine the underlying methods and principles of science; hence one historian's perceptive discussion of some of these works as 'metascientific'.[25] But the aim was as often imaginative, projective, and practical: to suggest not only how to think about the natural world, but to indicate modes of action and potential consequences in a period of unparalleled change.

No book of the nineteenth century took the reader on a more extraordinary imaginative journey than a small volume published in 1830, the chemist Humphry Davy's posthumous *Consolations in Travel, Or the Last Days of a Philosopher.* Davy, the most celebrated man of science of his day, saw his book as a final testament, an intervention in the debates that were tearing Britain apart as he lay on his deathbed. He supported parliamentary reform, but unlike Herschel and Knight condemned the diffusion of knowledge among the people as fundamentally mistaken. *Consolations* addresses an imagined clerisy of elite readers who could take society in a new direction. It is a series of dialogues on the progress of human life, the deep past of the earth, the relation between spirit and matter, the role of the man of knowledge, and the nature of a future state. In about 300 small pages, *Consolations* offers a dizzying range of perspectives on space and time: from the outer reaches of the solar system to the innermost structure of the atom, from the early history of life on earth to the future of the human race. It begins with a vision.

1

Fantastic Voyages

Humphry Davy's *Consolations in Travel*

> Reading Sir H. Davy's legacy 'Consolations in Travel'—what a sublime vision: I believe there is more than a vision in it; at least I hope so.
>
> —The geologist Gideon Mantell, journal entry for 5 March 1830

Alone on the moonlit steps of the Coliseum in Rome, gripped by 'a wildness and a kind of indefinite sensation', a visitor reflects on the decline of great civilizations. Suddenly the ruins vanish, the light of the moon blazes into splendour, and a deep melody fills the air. All the senses are heightened; all sense of personal identity is lost. A soft, clear voice from the centre of the light begins to speak. 'I am an intelligence somewhat superior to you, though there are millions of beings as much above me in power and in intellect as man is above the meanest and weakest reptile that crawls beneath his feet; yet something I can teach you: yield your mind wholly to the influence which I shall exert upon it, and you shall be undeceived in your views of the history of the world, and of the system you inhabit.'[1]

This scene is from the opening of *Consolations in Travel, Or the Last Days of a Philosopher,* a small book of dialogues that appeared in January 1830 as the final work of the chemist Humphry Davy. The vision continues as the narrator is guided by the spirit, the 'Genius', on a journey through the history of humanity, to be shown how civilization has proceeded from rude origins towards ever higher states of being. Midway through his passage across the centuries, the narrator sees again the Coliseum, but this time it is filled with a vast crowd watching the great gladiatorial combats, 'ornamented with all the spoils that the wealth of a world can give'. The works of Rome 'seemed more like the creation of a supernatural power than the work of human hands'.[2] But then all is lost through luxury and dissipation. Gradually there is a revival, and although the great military exploits of the ancient world have passed, those belonging to the learning and the arts begin to blossom again. The key to the lasting success of Christian civilization is the printing press. One individual, Johann Faust (who was generally credited in popular legend with the invention) has transformed the world:

> 'Now,' the Genius said, 'society has taken its modern and permanent aspect. Consider for a moment its relations to letters and to arms as contrasted with those of the ancient world.' I looked, and saw, that in the place of the rolls of papyrus libraries were now filled with books. 'Behold,' the Genius said, 'the printing press; by the invention of Faust the productions of genius are, as it were, made imperishable, capable of indefinite multiplication, and rendered an unalienable heritage of the human mind. By this art, apparently so humble, the progress of society is secured, and man is spared the humiliation of witnessing again scenes like those which followed the destruction of the Roman Empire.'[3]

The Genius goes on to explain the laws of history, how the races of northern Europe conquered the debilitated peoples of the Mediterranean, infusing them with the strength that ultimately led to the revival of learning in the modern era. It is a divine progression of happiness and spiritual enlightenment.

Do the laws of progress arise from physical organization, 'the machinery upon which thought and motion depend'? Or are mind and spirit specially created? To explain why neither of these explanations holds, the Genius offers a final journey beyond the veil of death. Raised high in a luminous column, the narrator sees a universe bursting with life, from the human-like creatures of Venus and Mars to the vast tentacles of the inhabitants of Saturn and the blazing intelligences of the sun. By the end, he travels as a beam of light into the realm of the angels. All these spheres are inhabited by ascending spiritual beings who have lived among the planets and are moving towards the sun. All is in progress, reborn spiritual natures rising towards infinite wisdom. The work of great men never disappears, but is preserved and built upon by future generations. Were the voyagers able to discern the fates of individual souls, they would discover the spirit that, having once taken the form in its earthly existence of the seventeenth-century natural philosopher and mathematician Isaac Newton, was 'now in a higher and better state of planetary existence drinking intellectual light from a purer source and approaching nearer to the infinite and divine Mind'.[4]

This was the first of six dialogues in *Consolations,* which discuss the early geological history of the earth; problems of generation and reproduction; the role of great men in history and in science; and the nature of time and immortality. A dialogue format and visionary use of fact offered opportunities for speculation on the widest questions of philosophy and belief, something that was becoming more difficult to do in scientific papers, which were increasingly focused on specific discoveries in the laboratory, field, and museum. *Consolations* responds to the crisis at the time of its publication, but does so by rejecting attempts to democratize the readership for science and addressing instead those selected spirits who could guide the nation to a higher destiny.

Philosophy in the Pocket

The significance of *Consolations* is apparent from *The Tenant of Wildfell Hall,* Anne Brontë's shockingly controversial novel of 1848.[5] Davy's work appears at a pivotal point in the plot, when the narrator, Gilbert Markham, discovers it on the table among the books of his beloved Helen Graham, whom he suspects is seeing another man. The volume is one he had not seen before in her collection, and in opening it Markham's worst fears seem realized, for it bears the signature of 'Frederick Lawrence'. This leads Markham to confront Graham. 'I felt disposed to dally with my victim like a cat. Shewing her the book that I still held in my hand, and pointing to the name on the fly leaf…I asked—Do you know that gentleman?' Lawrence, it turns out, is only her brother, but as a result of the incident Graham gives him a sheaf of papers from her private diary that reveal the terrible story of her marriage. The presence of *Consolations,* mistakenly interpreted by Markham as a sign of scandal and secret, turns out to be one more proof of the tragic depths of Helen Graham's character, and a compliment to her tastes as a reader.

Anne Brontë had long shown an interest in the book, and her choice is not coincidental. Like Mantell and many other readers, she studied *Consolations* as part of an engagement with contemporary religious and philosophical discussion in relation to the sciences. Reading the book in her father's parsonage in Haworth, Yorkshire, she wrote a manuscript dialogue of her own involving S and C, a 'Sceptic' and a 'Christian'. Here, she contrasted the Bible with other more 'tempting volumes', but also concluded with a long passage on geological progress drawn from *Consolations.* 'I know not any more sensitive or philosophical view,' the Christian in the dialogue states, than 'Humphry Davy's theory.'[6]

The physical form of *Consolations* had been precisely engineered for use in such enquiries. A small octavo, typical of polite books of reflection and thought for genteel readers, it was published by John Murray II, the great gentleman publisher of Albemarle Street in London's fashionable West End. Murray was well known for his fastidious attention to propriety—it was after all in his fireplace that Lord Byron's infamous autobiographical memoir was burned to ashes. But Murray also prided himself on having his finger on the pulse of intellectual fashion, which made a contemplative work by as dashing a man of science as Humphry Davy a valuable addition to his list.

Unlike the productions of the Society for the Diffusion of Useful Knowledge, with their cramped format and cheap paper, *Consolations* was a handsome production, though it was short and thus not particularly costly. It was printed by Thomas Davison of Whitefriars in London, who was known for the 'singular beauty and correctness of his works', including the use of a special ink that made the printed words on the page more sharply defined and easy to read. The results are evident in the final product: widely spaced lines, generous margins, cold-pressed paper, a small page size that was easy to hold in the hand. The original binding was simple paper-covered boards, which purchasers were generally expected to have rebound, typically in leather. The format had been used by Davison for another late work by Davy, the pseudonymous *Salmonia: or Days of Fly Fishing* of 1828, a small octavo that had been based in turn on the recent, hugely selling editions of Lord Byron's poems, also published by Murray and printed by Davison. At six shillings, *Consolations* was inexpensive enough to be bought by the upper middle classes, and it was held by the libraries of almost all provincial philosophical societies and mechanics' institutes. It was not really a cheap book, though, as it would have been possible to print the text in far fewer pages. Nor was it the larger size and shape characteristic of systematic, academic philosophy. *Consolations* sold exceptionally well, with seven editions in forty years, and a total sale in Britain alone of over eight thousand copies, not to mention its appearance in the author's collected works.[7] The book was widely circulated in the German-speaking world, and there were also early translations into Dutch and Swedish.[8]

The elegant format made *Consolations* instantly recognizable as an appropriate item on Helen Graham's fictional table. The book appears to have regularly served as a gift between the sexes, and many copies were owned by women. The kind of flighty reader that it is stereotypically understood to have attracted is shown in a print, 'Sketches from a Fashionable Conversatione', published a few years earlier (Plate 5). 'Oh!' says the woman on the left, 'I pant for Soul, Mind, Sentiment. I die for expression, delicacy, tenderness, and intellectuality.' No wonder that Gilbert Markham was suspicious about it being a present from a male competitor, even before he opened the covers to see the flyleaf. That title, *Consolations in Travel, Or the Last Days of a Philosopher,* needed no explanation to Brontë's readers, but it did indicate a sense of considerable complexity in relation to other works. *Consolations* had a clear reference to Boethius' Latin classic of the sixth century, the *Consolation of Philosophy,* with its spiritual guide; and also to various other works in the same tradition, such as Dante's *Divine Comedy.* Passages describing the history of civilization and of ancient life clearly relate to contemporary epic, and particularly to universal history. It also referred to projects by Davy's close friend Samuel Taylor Coleridge earlier in the century, especially *Consolations and Comforts from the Exercise of and Right Application of the Reason, the Imagination, and the Moral Feelings.*

But why consolations *in travel*? The reading of the book, even in a comfortable library or drawing room, was designed to call up the experience of reflective travel, being in size and shape a volume that might be taken on a journey. The small size of *Consolations* aimed at lightness and elegance. It was printed in what would have been known as a 'pocket' edition, which meant that readers could take it with them anywhere as a companion. As the pattern of exchanges in

Consolations makes clear, travel creates a distance from home that is essential to true reflection. England—the centre of the narrator's previous life—lacks the conditions (and not least the weather) vital to social stability and mental repose:

> Of all the climates of Europe, England seems to me most fitted for the activity of the mind, and the least suited to repose. The alterations of a climate so various and rapid, continually awake new sensations, and the changes in the sky from dryness to moisture, seem to keep the nervous system in a constant state of disturbance... in the changeful and tumultuous atmosphere of England, to be tranquil is a labour, and employment is necessary to ward off the attacks of ennui. The English as a nation is preeminently active, and the natives of no other country follow their objects with so much force, fire and constancy. And, as human powers are limited, there are few examples of very distinguished men living in this country to old age; they usually fail, droop and die before they have attained the period naturally marked for the end of human existence.[9]

London is a city of death: the narrator returns there only for the funeral of a friend, and leaves immediately afterwards because bodily and mental equilibrium can be retained only in tranquil landscapes and a contented society. Only in milder climates, or scenes of natural beauty, could 'the pains of disease' be replaced by 'repose and oblivion'.[10]

Each of the dialogues begins in a closely specified picturesque setting that creates mental repose and occasions philosophical reflection and discussion. The significance of the book's geography was highlighted in editions from 1851 onwards by wood-engraved illustrations at the front of each chapter; these were added by the hostess and naturalist Charlotte Murchison and other friends, based on drawings of their own continental tours. The territory, especially in the early chapters, was familiar to anyone who knew the writings of the enlightened deists of the eighteenth and early nineteenth centuries. Thus, the first dialogue, 'The Vision', takes place in the winter near the Coliseum in Rome, at the heart of classical civilization

Fig. 6. High above the earth, the Genius reveals the past and future of the universe to a 'friend of truth'. Vignette from the title page of an inexpensive edition of Constantin-François Volney, *The Ruins, A Survey of the Revolutions of Empires*, 5th ed. (London: Thomas Tegg, 1811).

and the setting for the opening of Edward Gibbon's notoriously anti-Christian but widely read *Decline and Fall of the Roman Empire*, and the form is modelled on Voltaire's *Micromegas*, which uses interplanetary travel to comment on the follies of mankind.

The most important model was the Comte de Volney's famous work of the late Enlightenment, *The Ruins: A Survey of the Revolutions of Empires*, first issued in Paris in 1791 and repeatedly republished in English during the following decades. Like *Consolations*, the *Ruins* begins with a 'profound reverie' above the ruins of a once-great city—in this case, Palmyra in the Syrian desert—and the narrator is guided by a Genius, a 'pale apparition, enveloped in an immense drapery' (Fig. 6). In Volney's vision, kingcraft and priestcraft will pass away, to be replaced by a faith unified around a God known not through Scripture or dogma, but the laws of nature. The *Ruins* appeared at the height of the revolution in France, which it sees as the culminating event of global history. Particularly with its appended *The Law of Nature*, the French revolutionary catechism, *Ruins* became a classic of freethinking religion, among the most widely circulated and reviled books of the early nineteenth century. The London printer Arthur Seale, known for his list of revolutionary and millenarian works, announced an edition of 1795 as 'sold by all political booksellers'.[11] Thomas Jefferson kept his involvement in another translation secret, presumably lest opponents use it to undermine his presidency of the United States.[12] In Mary Shelley's *Frankenstein* of 1818, the monster overhears the *Ruins* being read in a peasant's cottage in Switzerland, and learns of the rights of man, the origins of social inequality, and the catastrophe of organized religion. Percy Shelley, Mary's husband, had used the *Ruins* as the basis for his own arguments for the necessity of atheism in *Queen Mab: A Philosophical Poem*, which had been issued when the author was only eighteen. When the courts used the poem to prove that Shelley was an unsuitable father, cheap reprints from the pauper presses made it by far his most widely read work.[13]

As a travel companion, *Consolations* provided a striking counter-revolutionary alternative. The vision is set within the context of three English friends meeting to appreciate the ruins of Rome, each with contrasting views in contemporary religious debates: Onuphrio, a liberal aristocrat; Ambrosio, a Roman Catholic; and the narrator, Philalethes. They begin by discussing the relation between spiritual and material progress, an issue about which they differ passionately. Onuphrio, the sceptic, argues that the modern Rome will someday lie in the same broken state as its ancient counterpart. For the devout Ambrosio, in contrast, modern Rome will last until the end of days. Christianity is a final state of belief, the culmination of spiritual progress, and 'a creed fitted for the most enlightened state of the human mind and equally adapted to every climate and every people'. Onuphrio does not deny the value of religion, including Christianity: he acknowledges its importance in all its forms, from Hinduism to Islam, as 'belonging to the human mind in the same manner as instincts belong to the brute creation'.[14] But he does not believe that Christianity occupies any special position in relation to the material survival of civilization: the wood of the cross will decay as surely as the idols of eastern religions. This increasingly fraught exchange ends when Onuphrio and Ambrosio leave for a crowded soirée—precisely the kind of social distraction the narrator is eager to avoid. Staying behind, Philalethes experiences his fantastic vision. But here too, the contrast with Volney's *Ruins* and similar late Enlightenment works is evident, as the orientalist figure of an eastern sage is reinvented as a harp-like voice, ungendered and sexually ambiguous, adopted from the safely Christian traditions of the West.[15] This is the philosophy of the Grand Tour reconfigured for the new age.

During the following months, the protagonists move south along the Italian peninsula to the Kingdom of Naples, and from the summit of Vesuvius they discuss the meaning of the vision. Continuing their conversations, the third dialogue is set among the spectacular Greek temples of Paestum, where they meet a stranger, the Unknown, who extends the historical story into the distant past revealed by the new and potentially troubling science of geology. The fourth dialogue begins a fresh journey, this time to the Austrian Alps, which occupies the second half of the book. Philalethes, the narrator, is now dying. Together they tour the valleys of Illyria, and end up in a celebrated natural cavern, discussing the stages of life of the 'Proteus', a blind amphibian living in total darkness, which proves a convenient device for reintroducing the question of immortality (Fig. 7). The following evening the narrator asks the Unknown about his life and opinions, which leads to a long discourse on the role of the 'chemical philosopher', and a defence of the science of chemistry. On a final excursion in the eastern Alps, they travel to the harbour of Pola in Istria, where Philalethes, his physician, and the Unknown visit various Roman remains. The contrast between the beautifully preserved exterior of the amphitheatre, and its 'bare and naked' interior, leads them to consider the relations between time, matter, and progress.[16] They return to the themes raised initially in the vision in the Coliseum; but this time, guided by scientific wisdom of the Unknown, Philalethes can build on the lessons learned during the preceding travels.

Fig. 7. This illustration, later added to Davy's *Consolations*, shows specimens of the *Proteus anguinus* from caves in the Austrian Alps. The creature needed no light, could breathe both in the air and in water, and reproduced through an unknown mechanism. Humphry Davy, *Consolations in Travel, Or, the Last Days of a Philosopher*, 5th ed. (London: John Murray, 1851), 197.

Their meetings in the evocative ruins of ancient empires, between the Christian world and the East, open up general questions about the future of humanity. The reader follows a journey of enlightenment, traversing the borderland between life and death.

Celebrity and Reputation

What gave these reflections on time, rebirth, and regeneration both authority and poignancy was the fact that the 'last days' referred to in the subtitle of *Consolations* were not just those of the narrator, but also of the author, who had been the leading man of science of his generation. Readers, almost without exception, approached the book in relation to Humphry Davy's other writings, memories of his flamboyant evening lectures at the newly founded Royal Institution of Great Britain, and an ever-increasing deluge of details about his life.[17] Davy had been celebrated for discovering new elements, inventing the safety lamp, and rising from a relatively humble background in Cornwall. When he had tried to pay a penny to view the heavens through a telescope being shown by a street astronomer, the crowd made the exhibitor return the money, saying 'that's Sir Humphry'.[18]

During Davy's later years his daily movements were the subject of regular reports in the newspapers. Shortly after his death, his servant John Tobin published a memoir of the Italian and Austrian journeys, with intimate details about the composition and aims of *Consolations.* There were competing lives and letters, one by the chemist John Ayrton Paris, which was gossipy and indiscreet, the other by Davy's brother John, which was authoritative and admiring. By the end of the 1830s, when Davy's collected works appeared in ten volumes, readers had a wealth of published recollections, diaries, letters, and memoirs against which to judge the text of *Consolations.* Davy's story of spectacular self-improvement became a staple of the emerging literature of popularized science, in works ranging from the Society for the Diffusion of Useful Knowledge's *Pursuit of Knowledge under Difficulties* to Henry Mayhew's classic children's book, *The Wonders of Science, or, Young Humphry Davy; the Life of a Wonderful Boy: Written for Boys.*

Even in his early years of success, however, Davy had been mocked as a provincial Cornishman, insensitive to metropolitan ways and unable to adopt appropriate manners and dress. Davy was often portrayed as an effeminate 'fop' or 'dandy'. His presidency of the Royal Society was not a success, his attempts at reform stymied in part by his reputation for autocratic behaviour and arbitrary attempts to control the Society's warring factions. His marriage was a failure, his possible impotence openly discussed in public.[19] As he travelled to the Continent for a final journey in 1828, his reputation remained unsure. This can be seen from a newspaper clipping pasted in a contemporary scrapbook assembled by a fashion-conscious young woman. The extract highlights, not Davy's scientific contributions to humanity, but his inability to realize he was wearing five shirts at once.[20] For Davy himself, *Consolations* was a conscious effort to leave a final 'testament' that would recapture his early success and establish his reputation.

Within the main text of *Consolations,* the relation between the author's own voice and those of the various participants in the dialogues is remarkably complex. The first word is 'I', and the text is narrated very much in the first person; but the narrator's name ('Philalethes') is not revealed until the second dialogue. Clues about how the reader is to understand the book in relation to Davy's known views and reputation build up very gradually. In this respect, the text is unlike Boethius' *Consolation in Philosophy*, where from the start the author is awaiting execution. A fear of death motivates both works, but in Davy's *Consolations* that threat comes from within, from ennui and loss of inward energy; and when the text is at its most self-revelatory, in the penultimate dialogue on the 'Chemical Philosopher', the story is that of the Unknown, not the first-person narrator Philalethes.

If the main text only gradually encourages readers to read the book in autobiographical terms, the surrounding elements could scarcely be more explicit. Unlike *Salmonia*, which was authored pseudonymously 'By an Angler', *Consolations* has Davy's name emblazoned on the title page, and an opening note by John Davy explicitly asks readers to connect the text to his brother's life and recent death. In a short preface the author hopes that his own experiences on the borderlands of illness 'may be not altogether unprofitable to persons in perfect health'.[21] *Consolations* is thus entirely framed in terms of its readers' knowledge of Davy, although he is never identified with a particular character in the dialogues. In his widely read biography John Davy (like many reviewers) did equate his brother with the Unknown, although he also acknowledged that his brother repudiated this view. But the text frustrates such straightforward identifications at every turn. Not least, the dying philosopher in *Consolations* is Philalethes, and all the other characters state positions known to have been held by the author at different points in his life. There is no attempt at realistic speech, and many events in the book (such as an accident on a waterfall in the Austrian Alps) never happened. Reliance on a pre-existing sense of intimacy with the author is a widely used technique in the literature of romantic celebrity, appearing in works as disparate as Lord Byron's epic poem of self-revelation, *Don Juan*, and Coleridge's confessional *Biographia Literaria*. Readers and reviewers of such works were constantly teased into correlating specific episodes and opinions with those of the author, only to have those attempts frustrated or left uncertain. The literary form of *Consolations* thus allowed Davy both to draw on his public reputation and to distance himself from it: he is at once everywhere and nowhere in the text.[22]

The Lever of Dialogue

From Davy's perspective, these ambiguities had particular advantages in relation to writing about science. Scientific writing was undergoing a great transformation, with an increasing expectation that the author would communicate findings—usually in scientific periodicals—simply and plainly. Judged by this standard, Davy's publications during his early years in Bristol with the radical Thomas Beddoes were an embarrassment, not only for their potential materialism, but for their sheer imaginative energy. *Consolations* was an influential attempt to forge a new form for speculation in science. After all, the vision itself was a dream, literally (as Philalethes readily admits), 'a fiction', with ideas that were 'poetical' rather than 'philosophical': it could even be viewed as supernatural revelation (as the orthodox Ambrosio suggests), or as the working of a diseased imagination (as the sceptical Onuphrio points out).[23] A respected scientific and medical author in his own right, John Davy felt compelled to stress the 'unimpaired and unclouded' state of his brother's mental faculties at the time of writing it.[24] For him, the participants in the dialogues are ideal types, representing particular 'sentiments and doctrines' rather than realistic characters.

And that is the way the early dialogues work. The discussions on the vision expose contrasts between the different points of view, rehearsing debates about religion and reason from the late eighteenth century and in the wake of the French Revolution, very much along the lines of William Paley's *Natural Theology* and similar works. A typical confrontation occurs when Onuphrio and Ambrosio discuss the basic tenets of Christianity and the role of God as a creator:

> ONU.—Allowing the perfection of your moral scheme of religion and its fitness for the nature of man, I find it impossible to believe the primary doctrines on which this scheme is founded. You make the divine mind, the creator of infinite worlds, enter into the form of a man born of a virgin, you make the eternal and immortal God, the victim of shameful punishment and suffering death on the cross, recovering his life after three days and carrying his maimed and lacerated body in to the heaven of heavens.
>
> AMB.—You like all other sceptics make your own interpretations of the Scriptures and set up a standard for divine power in human reason. The infinite and eternal mind, as I said before, fits the doctrines of religion to the minds by which they are to be embraced. I see no improbability in the idea that an integrant part of his essence may have animated a human form; there can be no doubt that this belief constitutes the vital part of the religion. We know nothing of the generation of the human being in the ordinary course of nature; how absurd then to attempt to reason upon the acts of the divine mind! nor is there more difficulty in imagining the event of a divine conception than of a divine creation.[25]

Towards the end of the discussion, Philalethes moves closer to Ambrosio's position, but it is hard to see how there will be a resolution.

The introduction of the Unknown in the third dialogue decisively changes the dynamic of the dialogues. In contrast with the Genius, the Unknown is presented very much as a real person, wearing the marks of a pilgrim and a vial of chlorine to ward off the effects of malaria and to signal his status as a philosophical chemist. His history is outlined in the penultimate chapter. Born of humble parents, he had pursued a career of discovery with success. His pedigree is classical, in contrast with the alchemists of eastern legend:

> My life has not been unlike that of the ancient Greek sages. I have added some little to the quantity of human knowledge, and I have endeavoured to add something to the quantity of human happiness. In my early life I was a sceptic; I have informed you how I became a believer; and I constantly bless the Supreme Intelligence for the favour of some gleams of divine light which have been vouchsafed to me in this our state of darkness and doubt.[26]

In terms of dialogues of this period, the Unknown is a 'mentor', a character through whom the conversation could reach closure.[27] This is not because the Unknown comprehends everything directly, as the Genius had claimed to do up to a certain state of spiritual enlightenment. His views can be challenged, but in the end they provide a basis around which the views of the other speakers coalesce.

The Unknown's role in bottoming arguments is evident in discussions about the development of new forms of life, including humans, through the laws of matter. On the summit of Vesuvius, the Unknown introduces the wonders of the earth's past in a panoramic vision of progress, based on the latest findings of geology. As the earth gradually cooled from an original fluid state, shellfish, and corals of 'the first creation' appeared; as land formed, the first plants emerged. Higher forms followed, with fish being succeeded by reptiles, including monster saurians. The globe's crust was thin and the surface much hotter than today, 'in this state of things there was no order of events similar to the present'. These saurians were succeeded by higher and 'more perfect' creatures, including the mammoth and the megatherium. Finally man was created, and 'since that period there has been little alteration in the physical circumstances of our globe'.[28]

The Unknown's description of geological progress leads to a discussion of different theories of geology. The devout Ambrosio thinks a single creation more probable; but he is not a Scriptural literalist, and expresses satisfaction that Genesis is broadly in accordance with geology. The sceptical Onuphrio adopts the very different theory of the Scottish natural philosopher James Hutton, who had argued for a uniform cyclical order in the past. The Unknown comments that this view, taken to an extreme, would deny the fossil evidence for successive creations and imply an 'ancient and constant order of nature, only modified by existing laws'.[29] Oruphiro sees where this is going, and immediately draws back from defending the gradual development of new forms of life through the laws of matter. He would rather believe that fossils grew underground, or that they had been put there by God to test the faith of geologists:

> I am convinced;—I shall push my arguments no further, for I will not support the sophisms of that school, which suppose that living nature has undergone gradual changes by the effects of its irritabilities and appetencies; that the fish has in millions of generations ripened into the quadruped into the man; and that the system of life by its own inherent powers has fitted itself to the physical changes in the system of the universe. To this absurd, vague, atheistical doctrine, I prefer even the dream of plastic powers, or that other more modern dream, that the secondary strata were *created*, filled with remains as it were of animal life to confound the speculations of our geological reasoners.[30]

The Unknown responds by reinforcing the point that such views of the transmutation of species (which had been advocated by English poet and physician Erasmus Darwin and the French naturalist Jean-Baptiste Lamarck) are not part of modern science, but belong rather with folk legends and the views of uninformed believers in the literal truth of Scripture.

In a later dialogue, which takes place deep underground in the cave of the Maddalena at Adelsberg in the southern Alps, the narrator is relieved to hear such speculations refuted again by the Unknown. As Philalethes says:

> I am pleased with your views; they coincide with those I had formed at the time my imagination was employed upon the vision of the Colosaeum, which I repeated to you, and are not in opposition with the opinions that the cool judgment and sound and humble faith of Ambrosio have led me since to embrace. The doctrine of the *materialists* was always, even in my youth, a cold, heavy, dull and insupportable doctrine to me, and necessarily tending to atheism. When I had heard with disgust, in the dissecting rooms, the plan of the physiologist, of the gradual accretion of matter and its becoming endowed with irritability, ripening into sensibility and acquiring such organs as were necessary, by its own inherent forces, and at last rising into intellectual existence, a walk into the green fields or woods by the banks of rivers brought back my feelings from nature to God; I saw in all the powers of matter the instruments of the deity; the sunbeams, the awakened animation in forms prepared by divine intelligence to receive it; the insensate seed, the slumbering egg, which were to be vivified, appeared like the new born animal, works of a divine mind; I saw *love* as the creative principle in the material world, and this love only as a divine attribute.[31]

In the final pages, the sceptical voices fade away and Philalethes, building on the views of the Unknown, agrees that all philosophy must begin from a foundation of faith, and that this can be validated not only by studying God's works, but also by drawing parallels between the infinite mind of the divine and the mind of man. No further objections are presented to this conclusion, which has been achieved through a dynamic process of dialogue. The ambiguities and contradictions in individual positions lead, through the guidance of the Unknown, to a resolution. The book concludes with the reflections of Philalethes, who by this point knows he is dying. Astronomers have speculated on the extinction and birth of new systems of stars, and the same may also be true for the human spirit:

> It is, perhaps, rather a poetical than a philosophical idea, yet I cannot help forming the opinion, that genii or seraphic intelligences may inhabit these systems, and may be the ministers of the eternal mind, in producing changes in them similar to those which have taken place on the earth. Time is almost a human word and change entirely a human idea; in the system of nature we should rather say progress than change. The sun appears to sink in the ocean in darkness, but it rises in another hemisphere; the ruins of a city fall, but they are often used to form more magnificent structures as at Rome; but, even when they are destroyed, so as to produce only dust, nature asserts her empire over them, and the vegetable world rises in constant youth, and, in a period of annual successions, by the labours of man providing food, vitality and beauty upon the wrecks of monuments which were once raised for purposes of glory, but which are now applied to objects of utility.[32]

These issues had been debated for millennia, but *Consolations* attempts to provide a new context for their resolution. The Unknown's role as a man of science—a 'chemical philosopher'—is critical to his place as a mentor in the dynamics of the discussions in *Consolation*. Unlike the Genius, who is revealing phenomena not visible to ordinary mortals, the Unknown explains the findings of patient research from the laboratories of chemists, the fieldwork of geologists, and the observations of astronomers. His opinions can be discussed and debated, not as speculative propositions but as subjects open to further enquiry. Above all, he is a man of high moral character, who uncovers knowledge of utility for human life, and does not aim at transient popularity. The qualities required in the ideal seeker after scientific truth are so extensive, as the Unknown explains in the penultimate dialogue of *Consolations*, that they border on the improbable:

> I see you are smiling and think what I am saying in bad taste; yet notwithstanding, I will provoke your smiles still farther by saying a word or two on his other moral qualities. That he should be humble-minded, you will readily allow, and a diligent searcher after truth, and neither diverted from this great object by the love of transient glory or

> temporary popularity, looking rather to the opinion of ages, than to that of a day, and seeking to be remembered rather in the epochas of historians than in the columns of newspaper writers or journalists.[33]

The man of science should emulate 'the modern geometricians in the greatness of his views' (but implicitly, not their tendency towards scepticism), and imitate 'the ancient alchemists in industry and piety' (but not their shallow-minded delusions). It is through his qualities as a master of the sciences, both in their technical and moral dimensions, that the Unknown is able to move the narrative of *Consolations* towards closure. This use of a mentor is commonly found in the dialogue genres of this period, but it takes on a particular quality in *Consolations,* because of the grounding of the Unknown's moral authority in science. His interventions transform the dialogues into an idealized vision of the process of science as a model for the resolution of controversial issues of all kinds.

From Davy's perspective as an author, the dialogue form made it possible to recreate the experience of participating in scientific conversation, debate, and discovery. From his early days in Cornwall, Davy had kept a diary of his thoughts, experiences, and ideas in small notebooks, as a way of keeping different points of view in play as he worked through particular problems. As he focused increasingly on experiments (including his pioneering researches in the application of galvanism to chemistry), he continued to use notebooks in much the same way, so that discovery was a form of internal dialogue. His published experimental papers are in many ways conversations between various notebook passages, moving the reader towards conclusions based on diverse (and even initially confusing) evidence. The success of his scientific papers depended on a dialectical structure to build up what appeared to be an incontrovertible case, using what Davy called 'the lever of experiment' to give the illusion that nature was speaking, answering his queries directly.[34] *Consolations,* dealing with large and speculative questions about the future of humanity, could scarcely attain this demonstrative character, but moving between different positions towards a strong conclusion remains a central feature of the narrative. This was supremely important when an extraordinary range of opinions was being voiced in public on almost every conceivable question.

The Politics of Consolation

As indicated by the dismissal of journalists at the conclusion of *Consolation*, Davy had no patience with reviewers or the press more generally. One of the fly-fishing protagonists of *Salmonia* had raged against '*King Press*, whom I consider as the most capricious, depraved, and unprincipled tyrant that ever existed in England. Depraved—for it is to be bought by great wealth; capricious—because it sometimes follows, and sometimes forms, the voice of the lowest mob; and unprincipled, because, when its interests are concerned sets at defiance private feeling and private character, and neither regards their virtue, dignity, nor purity.'[35] Like Davy's other public interventions, *Salmonia* had had mixed reviews, the harshest being in the sharp-tongued Tory monthly *Blackwood's*, which found its characters lifeless, its settings improbable, and its style 'very, very trashy indeed'. The review did not mince words: 'silly', 'servile', 'mortally stupid', mechanically composed as if by 'an artificial pen' for a readership of young ladies. Most damning of all, the reviewer characterized the text as a 'patch-work composed of shreds of anniversary speeches before the Royal Society, articles in Philosophical Journals, and Lectures on Natural History to Mechanical Institutions'.[36]

With Davy's burial a recent event, *Consolations* was often reviewed in association with obituary notices. This inevitably tempered many of the reviews, although several picked up on the ambiguities of the author's later years. His prose was said to employ over-worn literary tropes, commonplace sentiments, and a hackneyed geography drawn from classical antiquity. As one weekly complained, the settings were banal and the literary machinery was like 'the device of the raree-show man, who pulls a string, and presents a new picture'. The elaborate apparatus, stately style, Latinate names of the protagonists: all continued to suggest a lack of manly confidence. If there was an implicit criticism, it was that the book was too bland, its settings too familiar, its religion too generalized. Evangelicals criticized its failure to advocate the religion of the heart: *Consolations* never mentioned the atonement of Christ crucified, and Davy was (as a correspondent to the *Wesleyan-Methodist Magazine* lamented) 'a sentimentalist, rather than an enlightened Christian'. Such readers regretted the use of a Roman Catholic as a mouthpiece for Christianity, but considered this as misguided broadmindedness, rather than a sign of secret heresy. Still, as the correspondent said, 'it turns, though often obscurely, to "feel after God"'.[37]

Although criticized for some of the same qualities as *Salmonia, Consolations* was generally well received. If Davy's wife Jane was successful, as she had hoped, in keeping knowledge of *Blackwood's* damning review from her husband, it still seems that lessons had been learned.[38] Not only were the locations in *Consolations* better motivated and the dialogues sometimes streaked with wit, but the stately language was better matched to the theme. The introduction of the Genius and the Unknown made it possible for the discussion to move beyond indeterminate controversy into a realm in which conclusions could be established, and certain speculations seen as potentially valid. As Davy told his old friend from Bristol, Thomas Poole, to whom the work would be dedicated, 'It contains the essence of my philosophical opinions, and some of my poetical reveries. It is like the "Salmonia", an amusement of my sickness; but "paulo majora canamus" [let us sing of greater things].' This was a reference not only to Virgil, but also to their common friend William Wordsworth, who had used this as an epigraph to his famous ode 'Intimations of Immortality'.

The refined use of dialogue in *Consolations* was undeniably a success. Given the skilled manipulation of the ambiguities of authorship, the dialogue form had the advantages of offering readers opportunities to place themselves in the text, to read from multiple perspectives, and to engage in a reflective process. Readers might, for example, have seen it as irredeemably heterodox, with its speculations on the transmigration of souls and a materialized version of a future state, with Jupiter and Saturn as way-stations to comets and the sun. Yet most accounts stress its piety. The novelist Sarah Harriet Burney, reading a copy loaned by a friend, noted that 'here are many fanciful and unwarranted ideas on the subject of the creation of this world, & the state of existence in the next: but on the whole, it is a most interesting work, and shews a mind anxious to discern the right, and well prepared to love and glorify its Creator'.[39] On this issue it was universally well-reviewed, the *Literary Gazette* praising the union of 'religious feeling and inculcation, with philosophical and natural remarks' and the weekly *Edinburgh Literary Journal* (in a dig at *Blackwood's*) for avoiding 'the affection and flippancy of the modern style of writing'.[40] The fashionable women's monthly *La Belle Assemblée* put it best: 'In this age of scepticism, how cheering, how grateful, how delightful it is to see a man so profoundly scientific, so deeply versed in the arcana of nature, as was Sir Humphry Davy, inculcating, upon clear, sound, *philosophical* principles, the sublime, the soothing, the ennobling doctrine of the soul's immortality!'[41]

Although *Consolations* was typically interpreted as a work of conventional piety that could be put into the hands of protestant Anglicans, one particularly despised minority, the English Roman Catholics, were eager to claim the book and its author. At the time Davy was dictating *Consolations* to his servant John Tobin, England was in the throes of debating the removal of legal restrictions on Catholics, who were unable to vote, hold public office, or attend the ancient universities. In letters to friends, Davy endorsed the emancipation movement, which is not unexpected given his advocacy of the Whigs and moderate reform. But his support seems to have gone well beyond this. In 1828, he was resident in Rome, attempting to recover his health, and engaged in long conversations with the Vice-Legate of Ravenna, Lavinio de' Medici Spada, a papal official and distinguished mineralogist. As the Anglo-Catholic *Dublin Review* later revealed, the principal subject of their conversations was religion. 'Sir Humphry was most anxious to be fully instructed in all the doctrines and practices of the Catholic Church,' the review noted, 'and . . . he expressed his determination, should the result of his enquiries answer his expectation, finally to embrace a religion, for the chief pastor of which [the Pope] he has expressed unequivocal respect and veneration.'[42]

As Davy hinted in another letter, he had become 'far more apostolic in my creed'.[43] Although there were rumours of a deathbed conversion, he was buried with protestant rites in Geneva, one of the strongholds of the Reformation. Yet what matters is not so much what Davy actually came to believe—but how easily his final work could be given a Catholic reading. The Birmingham-based priests who ran the *Catholic Magazine and Review* pointed out that the main advocate of Christianity in the dialogues, Ambrosio, is a Roman Catholic from England, who leads the narrator to a belief in the power of faith over reason, and who supports the use of relics and statues. Moreover, the Unknown—who was widely identified in the reviews as a version of Davy himself—believes in a 'universal' Christian Church, and even wears a rosary, which he kisses in memory of the Pope, 'a man, whose sanctity, firmness, meekness and benevolence are an honour to his church and to human nature'.[44]

Whatever its doctrinal merits, the perceived advantages of Roman Catholicism for the conduct of society are evident in Davy's late writings. *Salmonia* compares the happy peasants of the Alpine regions, secure in their faith and social position, with the agitated crowds of mechanics in England's industrializing north. It was fine if the lower orders (as in Scotland) were asking for enlightenment: but to push it upon them, as he saw Brougham and his supporters attempting to do in England, was a serious mistake, which could lead to dissatisfaction, unrest, and revolution.[45] Davy's opposition to spreading knowledge of chemistry and other sciences, is evident in a letter to his wife after she enthused about Brougham's tract on the *Objects, Advantages, and Pleasures of Science*:

> I become, however, every day more sceptical as to the use of making or endeavouring to make the people philosophers. Happiness is the great object of existence, and knowledge is good only so far as it promotes happiness; few persons ever attain the Socratic degree of knowledge to know their own *entire ignorance*, and scepticism and discontent are the usual *unripe fruits* of this tree—*the only fruits* which the people can gather; but I will say no more, knowing how unpopular my arguments will be; yet I could say *much*.[46]

Davy was here writing for an audience of one, his wife, Jane, and clearly recognized that the unpopularity of his arguments about the diffusion of knowledge started close to home. Lady Davy was not the only one who shared Brougham's views about scientific reading as a path to national redemption. The founder of the London Mechanics' Institution, George Birkbeck, singled out Davy for censure as one of 'our enemies'.[47] For his part, Brougham praised the writings of the great chemist, including *Consolations*, but believed that he had been unmanned and corrupted by the 'fashionable philosophy' of the metropolis.[48]

Working-class agitators, recalling that Davy had advocated the liberating potential of science in the revolutionary 1790s, went much further and condemned the conservative turn his politics had taken after his move to London. Writing from prison in 1821, the atheist publisher Richard Carlile had attempted to shame the great chemist back into the radical fold in an *Address to Men of Science:*

> The horror which was so lately expressed by the Emperor of Austria at the progress of Science, and at the revolution which Sir Humphry Davy had made in the science of Chemistry, is a specimen of that feeling which pervades all such men. This imbecile idiot quivered at an observation of his own physician about the state of his own constitution, and forbade him ever to use the word in his presence again! Yet it is by such men as this, that the inhabitants of Europe are held in a state of bondage and degradation! Will ye, Men of Science, continue to truckle before such animals?... Shame on you, if you can so far debase yourselves! Up, and play the man, boldly avow what your minds comprehend as natural truths; and all the venom of all the Despots and Priests on the face of the earth, shall fly before you as chaff before the wind.[49]

Such fiery polemics had little effect, other than underlining the dangers that the new sciences of matter might hold if placed in the balance against 'kingcraft' and 'priestcraft'.

The Christian Philosopher and the Fantasy of Empire

During the later decades of nineteenth century, the unified vision that readers had once found in *Consolations* began to fall apart. The work lived on through quotations in works for travellers, especially John Murray's famous handbook to southern Germany and the Austrian Empire.[50] These extracts generated a perspective on the work that prioritized scenic description and information for visitors over philosophical contemplation. The projected reader was a tourist on holiday rather than a contemplative traveller grappling with the complexities of matter and life. The illustrated editions from 1851 and after, with wood-engravings at the head of each chapter, brought out this concrete aspect of the work as a guidebook very strongly. *Consolations* was also read as an important contribution to the literature of empire. Its vision of universal history as racial conflict and the progress of European civilization became a commonplace, moving from speculation to assumption as the century progressed. Its explorations on the borderlands of Christian civilization were avidly read in colonial settings; one copy of the sixth Murray edition of 1853 is from the tiny library of Riwaka, a small town in the upper part of the south island of New Zealand.[51] Read as supporting industrial and racial hierarchies, it became a Victorian staple, the illustrated editions selling right up until the end of the century. Camille Flammarion, the great popularizer and fantasist of science, had it translated into French for the first time in 1868, where it went through many editions and had huge success. His reading of *Consolations* as imperial fantasy rendered it a harbinger of science fiction, in which the 'vision' was read as precursor of tales in which European exploration and conquest extended to other planets. Recent accounts characterize it as a pioneering account of alien life forms.[52]

Contemporary readers used *Consolations* to explore the place of science in understanding subjects as controversial as the relations between matter, vitality, and the afterlife. Perhaps the most striking outcome of its ambiguities of interpretation is that *Consolations* could be read as above politics and above sects. For a book to achieve this in 1830, at the peak of reform agitation, involved a remarkable exercise in balanced ambiguity. In this, genre was crucial. The book employed the unprecedented power of the 'celebrity author' within the context of the mentoring form of dialogues to bring potentially hazardous themes into a fresh—even consolatory—framework. The visionary 'sage' no longer needed to be an orientalized, eastern mystic in the mode of Volney's *Ruins,* but had become a scientific, Christian philosopher. Readers were invited to join a select group capable of penetrating deeper truths, to taste the fruits of knowledge.

Yet this new role for the investigator of nature was far from unambiguous. In many ways, *Consolations* was open to the same criticisms that the fashionable had levelled at Davy's clothes: a self-taught provincial, imitating the outward forms of established society, but unable to pull off the trick with confidence. The delicate format and typeface, too, carried at least some of the same sexual ambiguity that characterized Byron's poems. Was this a book for fops? For ladies? For Catholics? We are used to thinking of science in terms of philosophical rules of reasoning, but it is at least as much a matter of behaviour and tact. The proper conduct of the man of science was soon to become a public scandal.

2

The Economy of Intelligence

Charles Babbage's *Reflections on the Decline of Science in England*

A common stone meets with more ready patronage than a man of genius.

—[Richard Henry Horne], *Exposition of the False Medium and Barriers Excluding Men of Genius from the Public* (1833)

Whatever the failings of Davy in life, his early death on 29 May 1829 was widely seen as a tragedy for the world of science. With the fate of the sciences in Britain depending on individual endeavour, the loss of such a celebrated discoverer led to fears that the momentum achieved early in the century was petering out. Before his death, Davy had lamented the lack of interest in science among the current generation, especially among the aristocracy. As one of the participants in the dialogues of *Consolations* had said, 'there are very few persons who pursue science with true dignity; it is followed more as connected with objects of profit than those of fame'.[1]

Of all the innumerable recyclings of these words in print, the most notorious appeared in *Reflections on the Decline of Science in England, and on some of its Causes,* a short book written by the inventor, mathematician, and election manager Charles Babbage. The *Decline of Science* portrayed English science as moribund and corrupt, and looked to the

Continent, especially to France, for models of scientific reform. Appearing in May 1830, the book sold badly—it went through only one edition, probably of less than 750 copies—but it stirred up a storm of discussion, debate, and distress. Elizabeth Sabine was reportedly so upset by Babbage's accusations against her husband Edward, an astronomer and explorer, that she did not get out of bed for a month.[2] Babbage's remarkable polemic, idiosyncratic and intemperate, offered a vision of what science could do for human life, were it freed from the myriad abuses that (so the book claimed) dragged it down into the dirt.

From the early 1820s Babbage was best known as the projector of a calculating engine, which would mechanize the production of mathematical tables used in navigation and commerce (Fig. 8). Existing computers—humans—would be put out of work. The device employed a number of innovations first made in the printing trades, and also included a mechanical printer and automated typesetting, so as to remove every possibility of human intervention and error. At the time *Decline* was published, the Duke of Wellington's Tory government had provided munificent support of 7,500 pounds towards the construction of the device, making it one of the most significant engineering enterprises of the period. But the project was plagued by difficulties. Government finance had not always been so forthcoming, and relations between Babbage and the skilled craftsmen who turned his drawings into brass had been difficult. The reasons for this are not hard to find, for Babbage not only aimed to replace the mental labour of human computers, but also to obliterate the role of skilled labour in their construction. From Babbage's perspective, men such as the master craftsman Joseph Clement followed written instructions and drawings in an essentially mechanical way. They were human cogs in the machinery of what was effectively a highly developed printing establishment, turning written thoughts into a concrete

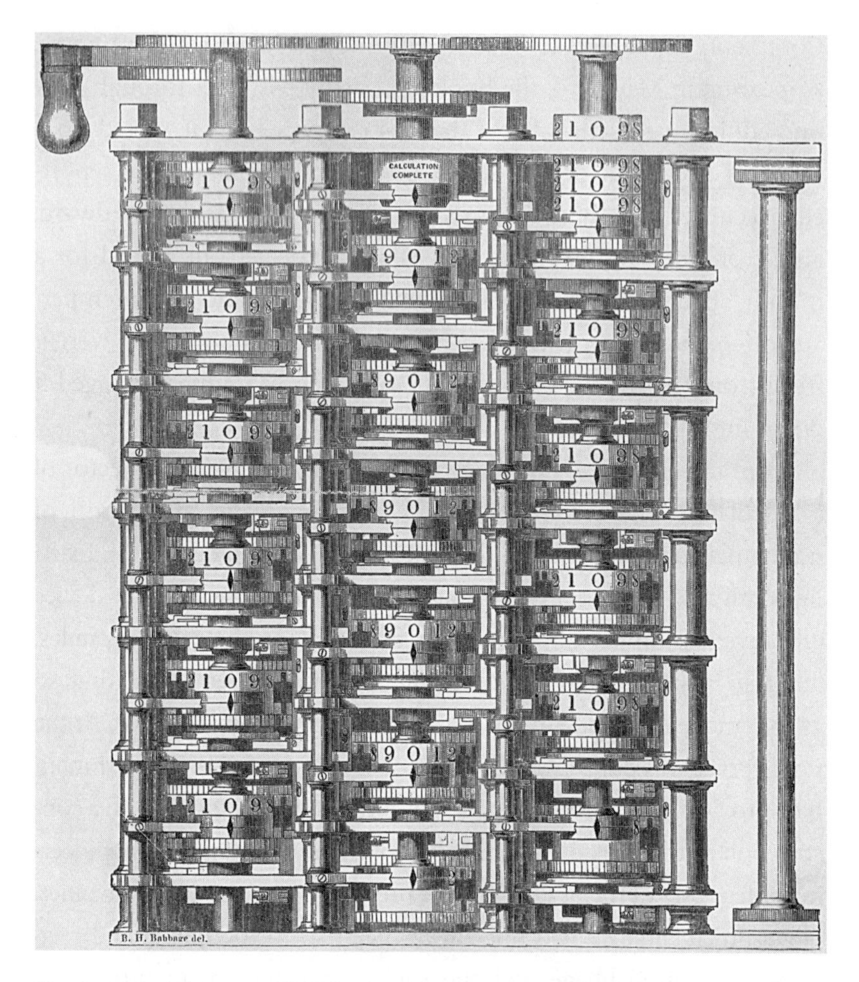

Fig. 8. Although none of Babbage's calculating machines were completed during his lifetime, this assembly could be used to demonstrate the basic principles involved in the operation of the difference engine. 'Small Portion of Mr Babbage's Difference Engine, No. 1 (Calculating Machine)', drawing by B. H. Babbage, from John Timbs, *Stories of Inventors and Discoverers in Science and the Useful Arts*, 2nd ed. (London: Kent & Co., 1863), following p. 142.

realized object: the printed mathematical tables that were to be the final output of the difference engine.[3]

Babbage's passion for calculating engines was part of his broader concern with what was known as the 'machinery question', which focused on the social, political, and intellectual impact of introducing steam power and mechanized production methods.[4] More than any of his contemporaries, Babbage believed that machines would ultimately reshape intellectual labour as fully as they were transforming the craft trades and manual work. For Babbage, the experience of constructing the engine underlined the dependence of the new economy on different forms of *intelligence*, ranging from the embodied skills of artisans such as Clement, through the rational management of industrial production, to the work of original discoverers such as Davy.[5] The introduction of machines, Babbage believed, would revolutionize intelligence at all these levels, and even the basic actions of reading and writing. The future of learning was tied to the future of machines.

Seen in this light, the coming of a machine-based economy raised profound questions about the conduct of science. Should research like that behind the calculating engine be supported by the state? What about scientific careers? The government gave pensions and salaries to leading figures in many walks of life, but only rarely to men of science. Yet if science became a paid occupation, its free pursuit might be threatened by a loss of independence. Should fellowship in the Royal Society, the oldest and most prestigious scientific society in the country, be limited to those who had published original scientific research? Should medical education stress mastery of the latest work in physiology and anatomy, or should physicians receive the classical education that remained the cornerstone of a gentleman's education?

From Babbage's perspective, these questions were tied to broader issues in the relation between machines and labour: the political economy of intellect in the new factory system. Much of the argument in *Decline* revolves around the issue of how research should be communicated in print, and how its results should be made available to various audiences. Among other issues, *Decline* intervened in the ongoing controversies about the role of reading, authorship, and the book trade in relation to British intellectual life, and it considers the issues of careers, government funding, and the organization of science from this perspective. Most significantly, Babbage was concerned to transform the role of authorship in science, by removing practices that he believed hampered initiative and obscured real merit. Reform in publication practices was seen as a key to achieving this.

Towards this end, *Decline* employs a particular persona for the author, one that is different from that to be found in Babbage's works on theology, mathematics, or political economy. In books like *Consolations*, the use of a dialogue form makes the futility of identifying the author with a single voice relatively obvious—although it is surprising how much time has been wasted trying to establish parallels between Davy's biography and the characters in the book. But in works such as *Decline*, and in fact for almost all non-fiction, we regularly assume that narrators are a transparent embodiment of their authors. In *Decline* we hear the voice of an angry accountant.

Something Rotten in the System

Like many of our own battles in intellectual life, the controversy over scientific reform was sparked by discussions over the fate of a library. At the Royal Society's annual meeting for 1829, the transfer of manuscript collections donated by the Earl of Arundel a century earlier was being debated. In a period when the boundaries of science were being narrowed and defined, everyone agreed that these literary and historical documents no longer had a place in a modern scientific library; but the terms of the transfer were hotly contested, with suspicions that valuable materials were being traded for unspecified books of unknown value, with unnamed individuals benefiting by the process. That would make it a 'book job',[6] yet another instance of corruption. The result was a fight at the meeting and volleys of letters to the newspapers.

This was the latest, and certainly not the largest, in a series of ongoing skirmishes about mismanagement in the Royal Society in the daily newspapers and the intellectual weeklies. Davy's successor as president, the engineer and antiquarian Davies Gilbert, was widely seen as amiable and mild, but completely ineffectual in combating corruption. The most reform-minded of the main dailies, such as the *Morning Chronicle*, had picked up the cry, as had the fiery radical medical weekly, the *Lancet*. In this paper war of attrition, even the smallest misstep by those involved in the management of scientific institutions produced a barrage of attacks in the press. In one of the last letters he wrote, Thomas Young, a leading natural philosopher and a member of the old guard, privately accused his opponents of being, not soldiers in the cause of reform but 'banditti' determined to bring down the scientific establishment.[7] As the 1820s drew to a close, even the casual reader of *The Times* and the *Morning Chronicle* knew that there was trouble in the world of science.

The turmoil went back at least to the death in 1820 of Sir Joseph Banks, a figure who, more than any other, epitomized the 'old regime' in science. Famed for his exploits on Captain Cook's early voyages in the Pacific, Banks had used his forty-year presidency of the Royal Society to promote a potent combination of natural history, antiquarianism, and agricultural improvement, bringing these to the service of Britain's rapidly expanding overseas empire.[8] Commanding a vast correspondence with everyone from prime ministers to commercial gardeners, he could speak as authoritatively about the cultivation of potatoes as about national trade policy. Banks, supreme patron of the natural sciences, in turn depended upon wider networks of patronage between the landed aristocracy, the royal family, and government ministers. From this perspective, the world of nature was a landed estate, well managed through an understanding of the practical arts, local traditions, and the sciences. Many of his followers looked to Banks's presidency of the Royal Society as a golden age. But

Banks was also hated in some quarters. The emergence of new, more socially diverse groups in science threatened existing webs of deference and patronage, and the vision of knowledge upon which they depended. Although he had supported the foundation of the Linnean Society in 1788 for the pursuit of his own favourite subject, natural history, from the early nineteenth century Banks and his allies jealously guarded the special role of the Royal Society, opposing those who wished to set up new groups for particular scientific specialities such as geology.

Shortly before Banks's death, a group of men of business, including bankers such as Francis Baily and wealthy 'grand amateurs' such as James South and John Herschel, began plotting a breakaway Astronomical Society that would pursue a reformed, mathematically refined, and secure version of the subject. The watchwords were speculation (to be avoided) and calculation (to be encouraged), with the ultimate aim of allowing the free flow of reliable information. From their perspective, the Royal Society was a bloated monopoly.[9] Those supporting the new society saw little to gain from affiliation with a body they increasingly saw as complacent, dormant, and dull. 'I see plainly,' Banks told Sir John Barrow of the Admiralty, 'that all these new-fangled associations will finally dismantle the Royal Society, and not leave the old lady a rag to cover her.'[10]

The triumphant success of the new scientific societies against such strong opposition marks the decay of older bonds of deference and obligation which grounded authority in social hierarchy rather than primarily in expertise. At the same time, groups such as the Geological Society and the Astronomical Society remained elite associations of independent individuals, mostly of gentlemanly status, and as such remained extensions of London's West End clubland. Babbage, a gentleman of independent income who cultivated the support of polite society, entirely endorsed this point of view.

Reform, whether in politics or in science, was, from this perspective, best carried out by men of independent judgment.

After Banks's death the forces of change could no longer be kept at bay, and in agreeing to be put forward as his successor as Royal Society president, Davy had realized that managing competing factions would be his main task. Initially he placated the reformers, placing them on Council and awarding them key positions. From the viewpoint of the reformers, however, Davy was weak, vacillating, and too prone to bow to claims of rank and privilege; as we have seen, he was openly criticized in the conservative press as a fop. His initial overtures towards change were widely considered to have petered out long before illness forced his resignation in 1827.[11] But even before then, demands for scientific reform became increasingly vocal. These combined with older grievances: Oxford and Cambridge had only partially reformed, and arrangements for medical education in England were increasingly seen by radicals like Thomas Wakley as outdated, taking little account of current research on the Continent or the need for practical training.[12]

From the reformers' perspective, the catastrophic weakness of the sciences was demonstrated in the summer of 1828, when Parliament abolished the Board of Longitude, which had been founded a century before to encourage the solution of the problem of finding longitude at sea, a vital issue for a nation that depended on accurate long-distance navigation. The Board was a soft target for a government struggling to make savings and sensitive to accusations of awarding sinecures. With John Harrison's chronometers in regular use, enabling ships to locate themselves with considerable accuracy, it was easy to argue that the original aim of the Board had been achieved. Among its remaining responsibilities was the *Nautical Almanack*, a key compendium of information for celestial navigation, although this too had been subject to sustained criticism for inaccuracy and the arrangements for its publication. The *Almanack* was reorganized under a new committee, mainly as a way of dodging further censure.[13] The reformers viewed closure of the Board as a terrible blow, at a time when they were arguing for increases in posts and funding.

These longstanding grievances were coming to a head just as the issue of the Arundel manuscripts was being discussed at the Royal Society. In a letter to *The Times* a few days after the annual meeting, the astronomer James South, one of the leaders of the group clamouring for reform, recalled talking with Davy about this issue just before his final departure for Italy. Davy showed him what South described as a thirty- or forty-page manuscript 'on the decline of science in Great Britain, and on the causes which have led to it'. South called for Davy's tract to be published—although in fact it seems to have been destroyed. 'The pages that I saw,' South recalled, 'were dictated by the purest love of science, were stamped by the consciousness of the decline of it in his native country, were written with uncommon energy, and were founded in truth: they were creditable to his head, and honourable to his heart.'[14] Davy had been a disaster as a practical manager of the future of science at the Royal Society, but reconfigured posthumously as a figurehead for reform he could scarcely be bettered.

A Public Accounting

In contrast to the sniper-fire of anonymous newspaper letters, Babbage aimed for full-scale warfare. Events at the Royal Society in 1829 appear to have decided his course. At the meeting he argued that the agreement for disposing of the Arundel manuscripts was '*unwise*', '*improvident* and *unbusinesslike*', as the terms of the exchange were unclear. As Babbage saw it, a better solution would be to sell the manuscripts on the open market and then use the money to buy the books really needed for the Society's library.[15] Of course, this would mean that the manuscripts were no longer available for consultation, but that was a matter outside the Royal Society's concern. A cash transaction, conducted in public, established value in a way that a private negotiation never could.

This issue of transparency, introducing 'businesslike' mechanisms of exchange in areas previously dominated by private interest, was at the heart of the much larger issues discussed in *Decline*. Indeed, Babbage's decision to write a more extended signed work arose from his conviction that the reform debate suffered from problems created by longstanding traditions of anonymous and pseudonymous writing in periodical controversies. Most articles and editorial letters in newspapers, literary journals, and other public forums were unsigned. The *Edinburgh Journal of Science* tried to use this as the dividing line between the 'declinists' and their opponents:

> The present dispute presents to us the rare and almost ridiculous feature, never before exhibited in literary controversy, of *all* the combatants on one side coming openly forward with their names, and pledging their public characters for the truth of their facts and statements; while *all* those on the other side are anonymous assailants, fighting behind black crape, or discharging their missiles from the masked batteries of magazines and reviews.[16]

Davy, Herschel, Babbage, South, and other active researchers on the one side; 'Socius', 'Paul Pry', and 'One of the 680 F.R.S.' on the other. But this was hardly the full story, as 'Job Hater', 'Censor', 'A Constant Reader', and many others in the declinist camp were presumably not writing under their own names. Ironically, the most significant anonymous contributor to the debate was the natural philosopher David Brewster, who never signed any of his articles on the topic, including the one just quoted from the *Edinburgh Journal of Science*.[17]

Babbage, who did put his name to what he wrote, agreed that unsigned newspaper attacks had led to a lack of clarity about whose interests were actually at stake, even in the 'diminutive world' of science.[18] He initially planned to write a pamphlet which would give the names of the offending parties, with his name on it, and towards that end began to gather ammunition for an attack on corruption in English science. In order to demonstrate the covert way that awards were made by the Royal Society, Babbage asked the Society's secretary for the detailed terms on which one of the Royal Medals had been just been awarded; but he was turned down. He asked for a free copy of the published observations of the Greenwich Observatory, as it appeared these were in danger of being pulped after not being made adequately available to interested individuals; but this request was refused. He petitioned for access to the council minutes about an important appointment; but this was denied.[19]

The purpose, as Babbage explained to his friend Herschel, was 'to cut up the coteries' in the Royal Society, and 'explore some of the abuses and leave the thing to be corrected by public opinion'.[20] By early March the pamphlet had become a short book. It was published by Benjamin Fellowes, who had issued several of Babbage's previous works, and John Booth of Duke Street, proprietor of a circulating library, as an octavo of 228 pages, priced at a substantial seven shillings and six pence, and printed by the rising firm of Richard Clay with wide margins and neatly set type. The general appearance, as well as the tone, was that of a polemical political pamphlet. The 'Preface', for example, carefully explained why the book included attacks on named individuals:

> With respect to the cry against personality, which has been lately set up to prevent all inquiry into matters of scientific misgovernment, a few words will suffice.
>
> I feel as strongly as any one, not merely the impropriety, but the injustice of introducing private character into such discussions. There is, however, a maxim too well established to need any comment of mine. The public character of every public servant is legitimate subject of discussion, and his fitness or unfitness for office may be fairly canvassed by any person. Those whose too sensitive feelings shrink from such an ordeal,

> have no right to accept the emoluments of office, for they know that it is the condition to which all must submit who are paid from the public purse.[21]

As Babbage had said in his letter to Herschel, he 'determined not to spare the meanness of these councils and principles to which I might have been the victim. Had I been supported as I ought to have been my feelings would not have been so strong and my friends would have had some right as well as success in restraining my pen.' The work, he admitted, 'will be a receipt in full for the amount of injuries I have received and like a bill once paid I shall forget them and my anger together'.[22]

By expanding his work into a book, Babbage was able to extend his criticisms to detail the full range of problems at the Royal Society and to show it was not an isolated case. Surveying the range of independent specialist societies, he pointed to the potential for financial irregularities at the Horticultural and Zoological societies, and the disastrous situation of the Medico-Botanical Society, which had been founded in 1821 to encourage understanding of the medicinal uses of plants:

> The Medico-Botanical Society suddenly claimed the attention of the public; its pretensions were great—its assurance unbounded. It speedily became distinguished, not by its publications or discoveries, but by the number of princes it enrolled in its list. It is needless now to expose the extent of its short-lived quackery; but the evil deeds of that institution will long remain in the impression they have contributed to confirm throughout Europe, of the character of our scientific establishments. . . . As with a singular inversion of principle, the society contrived to render *expulsion* the highest *honour* it could confer; so it remains for it to exemplify, in *suicide*, the sublimest *virtue* of which it is capable.[23]

In fact the Medico-Botanical Society did not expire, although it never fully recovered from the uproar surrounding the expulsion of Robert Brown, the most celebrated botanist in the country.

From Babbage's perspective, publication of a separate work had the signal advantage of moving the controversy about the disastrous state of science beyond letters to newspaper editors, and into the quarterlies and leader columns. Independently wealthy and educated at Cambridge, Babbage had a wide acquaintance with metropolitan literary and political circles, which he used, through his famous evening parties in his house at Dorset Street, to raise the position of science in national life. To publish a book-length work was to acknowledge the dispute as a matter of lasting significance that would be recorded in other books and pamphlets, as well as in reviews that were regularly bound and preserved in libraries. *Decline* gave editors a 'peg' on which to hang substantial discussions, and an editorial in *The Times* recognized that 'there must be something rotten in the system' and called for a response from those Babbage had accused.[24] A host of pamphlets, books, and reviews on both sides were rushed into print in the following months, with titles such as *Charges against the President and Councils of the Royal Society* and *Science without a Head.*[25]

By bringing together all the criticisms that had been offered up piecemeal in newspaper letters and comments at meetings, *Decline* polarized the controversy. It was easy to read this devastating attack as the manifesto of a 'declinist party', a view encouraged by a long essay in the *Quarterly Review*, which supported and extended almost all of the book's arguments. This was widely known to be from the pen of David Brewster, and proved an important salvo in his ongoing campaign to raise the status of the scientific practitioner. 'The decline of science,' he told Babbage, 'was "the most heartbreaking subject that I know"', and he gave examples from his own experience in Scotland.[26] However, many who had favoured specific aspects of scientific reform refused to be drawn into what they saw as a partisan Babbage–Brewster axis. William Whewell, a rising star at the University of Cambridge, was outraged, and listed the results of a sustained campaign to raise the standard of science at the ancient universities in an article in the *British Critic*.[27] All this turbulence, he worried, muddied the waters and damaged prospects for reform. One of the schemes to improve the status of science, instigated by Brewster and endorsed in *Decline*, was to have a national gathering of scientific men along the lines of a similar union in the German lands. However, the fracas over the Babbage's book created obstacles that almost scuppered what became the British Association for the Advancement of Science. It took all the tact of those organizing the new body to ensure that the first meeting in York went ahead in the summer of 1831.[28]

Make them Writhe

It is easy to read *Decline* as truculent and vindictive, but the author did not see things that way, believing instead that behind personal grievances were larger abuses that could be logically demonstrated. Although, at a personal level, Babbage felt he had plenty to complain about, his book argued that the issue was one of principle. Like John Wade's populist polemic *The Black Book, or Corruption Unmasked! Being an Account of Persons, Places, and Sinecures* of 1820–3, this was a thorough accounting of corruption, from a perspective of Whig opposition.

The astronomer Edward Sabine, for example, was accused in *Decline* of cooking vital pendulum measurements and illegitimately accepting 1,000 pounds from the Admiralty for results that both parties must have known were too good to be true. *Decline* begins the account with heavy irony, noting that the agreement of the observations with one another was as remarkable as it was unexpected, and 'creditable to one who had used some of the instruments for the first time in his life'.[29] It then detailed how the work had been praised in the highest terms (the Royal Society had awarded a Copley Medal, its greatest honour), before giving the simple but specific reasons why the measurements were wrong. The ultimate fault was with the Admiralty officers who knew too little science to recognize that Sabine was a charlatan:

> Enough also has appeared to prove, that the conduct of the Admiralty in appointing that gentlemen one of their scientific advisers, was, under the peculiar circumstances, at least, unadvised. They have thus lent, as far as they could, the weight of their authority to support observations which are now found to be erroneous. They have thus held up for imitation observations which may induce hundreds of meritorious officers to throw aside their instruments, in the despair of ever approaching a standard which is since admitted to be imaginary; and they have ratified the doctrine, for I am not aware their official adviser has ever even modified it, that diminutive instruments are equal almost to the largest.[30]

Unsurprisingly, Sabine refused to meet the author at dinner, and replied icily to the charges in a paper in the *Philosophical Magazine*.[31] *Decline* anticipated such reactions, but argued that they were inappropriate, as long as the justice of the case was clear. Readers, and particularly the individuals attacked, should operate in the same rational way that was demonstrated by the author's own practice, by setting out a case under their own names. As holders of paid positions, or as leaders responsible for particular bodies, men like Sabine were expected to respond to the force of evidence and the logic of argument. A true philosopher, faced by accusations of corruption or forgery, would remain calm. Only those with something to hide would react badly; openness was the sign of honesty. 'There are few circumstances which so strongly distinguish the philosopher, as the calmness with which he can reply to criticisms he may think undeservedly severe.'[32] Like a true accountant, the narrator of *Decline* directed his fire at those who worked for pay, whether for the state, for a public company, or for a learned society. The book named names, and claimed not to hide criticisms of individuals under general condemnation.

In this way, *Decline* developed a supremely rational and highly unusual model of authorship and reading. The exchange of truth was assumed to be beyond controversy or interpretation. As Babbage later said during a parliamentary campaign, in philosophy as in natural philosophy, 'it was his rule first to be careful in the obtaining and sifting of facts; and in the next place of reasoning on them fearless of the legitimate consequences of his reasoning'.[33] Babbage had no faith in public opinion in its current unenlightened state, and marvelled at the extraordinary range of ways that readers approached his books. His hope was to bring his observations on this subject 'to a kind of demonstration', and towards that end he pasted printed comments from readers about *Decline* and his other books in a large scrapbook, with views for and against in parallel columns. This was not out of personal vanity, but to subject opinion (as his friend and fellow advocate of political economy Harriet Martineau explained) to the rigours of the balance sheet.[34]

Reviewers pointed out that the structure of *Decline* bore all the inexorable logic of a mathematical proof. Even friends laughed at Babbage's tendency to reduce everything to mathematics. The geologist Charles Lyell wrote in his journal that Babbage 'unconsciously jokes and reasons in high mathematics, talks of the "algebraic equation" of such a one's character, in regard to the truth of his stories, &c.'.[35] The book did not, however, attempt to demonstrate that English science was in decline; it took this as a given (citing a few authorities, notably Davy and Herschel) and moved on from there. Picking up on this, the *New Monthly Magazine* sarcastically noted, 'we strongly suspect that the learned author meant to treat the subject mathematically, by assuming the imaginary quantity, and then working out his inverse reflections, in the way he would an algebraic equation'.[36]

The mathematical trajectory of decline began with faults in the contemporary educational system. Praising the recent introduction of optional lectures on chemistry, geology, botany, and other sciences in the universities, the book made the radical suggestion that these subjects should be examinable, with success rewarded as part of a degree course. Young men were currently discouraged from cultivating science after leaving university because science—unlike medicine, the clergy, law, and the military—was not 'a distinct profession', with status and pay. Gentlemen from Cambridge and Oxford (like Babbage himself) might be willing to undertake the long train of study required for mastery of mathematics or geology, but needed situations appropriate to their status. The profession that attracted the highest talent in England—the law—appeared to have least immediate connection with science. On the other hand, the widespread pursuit of science by the clergy was inappropriate, owing to dilettantish amateurism and the wrong kind of training of the mind. The answer, as in the case of the universities, was a professional status. For the scholar aspiring to pursue science, *Decline* pointed out, 'There are no situations in the state; there is no position in society to which hope can point, to cheer him in his laborious path.'[37] Suitable positions could come from government, the book argued, but in the current state of affairs this was unlikely to happen. Experience suggested that the governing classes knew little of science, and those they consulted (such as Sabine) hardly commanded the respect of active practitioners in the field and laboratory.

Decline then attacked the most glaring symptom of this failure of governance, the leadership and organization of the Royal Society. It began by recounting the origins of specialist scientific societies, showing with heavy irony how the Royal Society had cultivated an attitude of 'discouragement and opposition' as the best way to make these competing bodies thrive. 'Finding their first attempts so eminently successful,' Babbage wrote, 'they redoubled the severity of their persecution, and the result surpassed even their wildest anticipations.'[38] The dangers threatening the future success of groups such as the Geological and Astronomical societies were suggested by what had happened to the Royal Society in its years of decay. All scientific societies needed to keep regular public accounts, maintain an inclusive leadership, and avoid presidents for life. The character of such institutions in England as associations of individuals, not dependant on government finance, was a source of strength, but was also open to abuse. For those who did not actively pursue science, Babbage cynically suggested, the principal benefit of membership was listing the appropriate letters after one's name—for anyone who could pay the entrance fee and get a few friends to sign a certificate could join. 'Thus,' he pointed out, 'those who are ambitious of scientific distinction, may, according to their fancy, render their name a kind of comet, carrying with it a tail of upwards of forty letters.'[39]

The most celebrated of those letters was 'F.R.S.', Fellow of the Royal Society; but that currency had been devalued, not only because membership was too open but because of corruption and jobbery. By far the longest chapter of *Decline*, occupying over half the text, is a meticulous accounting of the Society's failings. If a society such as the Astronomical showed what could go right when an institution operated according to the laws of political economy, the Royal showed what could go wrong.

As presented to readers of *Decline*, reform of the Royal Society served as a model of what needed to be done more generally in politics and the emerging industrial economy. Scientific education, for example, should aim not only to produce a cadre of specialists, but also to introduce rational modes of thought among the governing classes. Without this, even reforms in the 'little world' of science were bound to fail. For example, the book's suggestions that the government should give special prizes to men of science, and paid posts to those carrying out independent trains of research, depended upon a sufficiently large proportion of those in power having received an education based on the principles of scientific reasoning, rather than the combination of classics and theology that most of them currently studied at Oxford and Cambridge.

The ideals of reason and transparency are most evident when *Decline* focuses on issues of accuracy and fraud in observation. Amid the resolute accounting of the preceding pages, this may seem a curious philosophical interlude. It is here, however, that the book sets out most clearly the foundations of a perspective relevant not only to science but also to politics and to managing the economy. Why did people sometimes disagree, or see things in different ways? One way of answering these questions pointed towards the possibility of calibrating observers. In *Decline* Babbage recalled tests that he made with a friend, using a mechanism intended to mark out minute divisions of time. They found that the difficulty of registering the twentieth part of a second depended on a variety of identifiable factors, including the time it took for the transmission of the will to stop the clock and the time employed in compressing the fingers. Different people had different relations to the machine, which could be rigorously compared.

A more disturbing reason for disagreement was fraud. This included hoaxes ('intended to last for a time and then be discovered, to the ridicule of those who have credited it'); forging ('wishing to acquire a reputation for science, records observations which he has never made'); trimming ('clipping off little bits here and there from those observations which differ most in excess from the mean'); and cooking ('to give to ordinary observations the appearance and character of those of the highest degree of accuracy').[40] The claim is that outright hoaxes and forgery are relatively rare, but that constant vigilance was required to spot the lesser known but more pervasive forms of fraud. The taxonomy of fraud in *Decline* provided a tool whereby science could assume its rightful status as a completely open enterprise, comprehensible not just to adepts and experts, but to 'ordinary understandings' more likely to be fooled or misled. 'He who can see portions of matter beyond the ken of the rest of his species,' Babbage wrote, 'confers an obligation on them, by recording what he sees; but their knowledge depends both on his testimony and on his judgment. He who contrives a method of rendering such atoms visible to ordinary observers, communicates to mankind an instrument of discovery, and stamps his own observations with a character, alike independent of testimony or of judgment.'[41] This ability to make the truth universally visible was the core of what it meant to be a true man of science: 'The character of an observer, as of a woman,' the book noted, 'if doubted is destroyed.'[42] That is what had happened to the 'old lady', the Royal Society. It was not that its leaders had produced false science, but they had produced an institution based on false principles, and *Decline* offered positive suggestions for its improvement and other changes in the organization of science. Freed of human subjectivity and foibles, the pursuit of knowledge would be manly and secure, a suitable model for political action. It is with this issue of character in mind that the book concluded with eulogies of Davy and the chemist William Hyde Wollaston, regarded as the greatest recent losses to English science.[43]

The stress on openness and character is one among many indications that *Decline* was a prequel to Babbage's work on political economy, *On the Economy of Machinery and Manufactures,* which went through three large editions in 1832 and a final one in 1835, as well as many foreign translations.[44] This book aimed to accomplish broader political work in education by confronting the problems created by the public's lack of access to engineering and industrial information, and arguing that the apprentice-based oral culture of Britain's workshops and factories led to waste and inefficiency. Unlike *Decline,* it was a cheap book (or as cheap as books could be given government taxes and the state of the print trade) in a small format. It made available a wide range of hitherto secret practical knowledge. The role of the narrator in the *Economy of Machinery* is not that of *Decline*'s accountant checking the books of a dodgy firm, but the bringer of light to dark places, describing clearly and simply processes that were unknown to the public at large. Among the most telling disclosures of trade secrets were the costs behind books themselves. Babbage exposed the corrupt and conservative nature of the book trade as thoroughly as he had revealed the state of the Royal Society, but claimed that both could be reformed. The book's intended publisher, Benjamin Fellowes (who had published *Decline*), was so outraged by revelations about the costs and techniques of book production that Babbage was forced to look to Charles Knight, publisher for the Society for the Diffusion of Useful Knowledge, who shared many of Babbage's attitudes towards the free flow of knowledge.

The most significant long-term effect of *Decline* was to open a general discussion of the political economy of intellectual life. Many readers came to believe that the coming machine economy would depend on learning of many different kinds, for which traditional patronage from wealthy individuals did not appear sufficient or well directed. Nor was reliance on the free market the answer, for this meant that booksellers catered to the lowest common denominator of taste in reading. There were almost no subsidies for publishing learned works compared with what was done in Paris, Göttingen, and St Petersburg. In 1831, a pamphlet by the archaeologist James Milligen claimed that in dealing with the physical sciences the book had tackled an area in which government support was relatively secure, compared at least with what was done for history, languages, literature, and other 'Intellectual Sciences'. The Royal Society might need reforming, but its counterpart in the study of the past, the Society of Antiquaries, had 'slumbered for years' and the British Museum was badly managed and viciously corrupt.[45] The architect William Wilkins agreed, arguing that patronage of the arts was ill-coordinated and miserly, based on his experience in designing the National Gallery of Art on Trafalgar Square.[46] These views were brought together in Edward Bulwer-Lytton's magisterial *England and the English* of 1833, which made a thoroughgoing case for extending political reforms to the institutions of knowledge.[47]

These failings had consequences for Britain's influence as an imperial power. Readers of *Decline* were made thoroughly aware that outdated practices at the Admiralty, the Royal Society, and other scientific bodies affected global commerce. The *Asiatic Journal*, a London-based clearing-house for readers interested in affairs in India and China, published a long review drawing attention to the significance of the declinist case.[48] The argument was renewed in the following decade in *Thoughts on the Degradation of Science in England*, written under the pseudonym 'F.R.S.' by the topographer Colonel William Martin Leake, who had served in Mediterranean campaigns and was a passionate advocate of Greek independence. His book looked back to the famous controversy over Babbage's book to warn that ignorance of subjects from climate and ancient languages to antiquities and naval architecture was already hampering military operations. The disasters of the British retreat from Kabul in 1842 during the First Afghan War—in which tens of thousands died—could have been avoided had science spoken to power with a proper understanding of geography and history.[49] The contest for intellectual ascendency had consequences that were not confined to Europe, but involved the future of empire.

A Vision of Politics

Cheered on by 5,000 people on Islington Green soon after publication of the *Economy of Machinery* in 1832, Babbage accepted nomination as a Whig parliamentary candidate for Finsbury at the first election following the Reform Act. He proclaimed himself 'since the dawn of life a Reformer', and supported more frequent elections, secret ballots (although he later changed his views on this), and the reduction of government expenditure, especially in relation to the established Church. He argued for lowering taxes on corn, on window glass, and above all on the spread of knowledge. Here his reputation as a man of knowledge was vital, for he could claim to understand the expenses of introducing patents, and the limits on the spread of science among ordinary people produced by taxes on paper and on newspapers. 'Give me but knowledge for them,' Babbage said, 'and I fear nothing.'[50]

The weekly *Mechanics' Magazine* was ecstatic, and urged newly enfranchised voters to turn out in force:

> Political matters do not, generally speaking, come within the province of this journal, but when the interests of science happen to be mixed up with them why should we be silent? We look for a great deal of good to science, as well as to every other important interest of the country, from the return to Parliament of a gentleman of Mr. Babbage's eminence in the scientific world and, therefore, we take the liberty to say to every elector of Finsbury who is a reader of this journal and a friend to the objects it has specially in view—Go and *vote for* Mr. *Babbage.* If you are an inventor, whom the iniquitous and oppressive tax on patents shuts out from the field of fair competition, and are desirous of seeing that tax removed—Go and *vote for* Mr. *Babbage.* If you are a manufacturer, harassed and obstructed in your operations by fiscal regulations—and would see industry as free as the air you breathe—Go and *vote for* Mr. *Babbage.* If you are a mechanic, depending for your daily bread on a constant and steady demand for the products of your skill, and are as alive as you ought to be to the influence of free trade on your fortunes—Go and *vote for* Mr. *Babbage.* If, in fine, you are a lover of science for its own sake alone, and would desire to see science honoured in those who most adorn it—Meet us to-day on Islington-green and *vote for* Mr. *Babbage.*[51]

In the end, the Tories triumphed in Finsbury. The reform vote split between three candidates, including the late-entering medical journalist Thomas Wakley, whose advocacy of immediately freeing slaves in the colonies appealed to religious dissenters, and whose opinions (and theatrical tactics) were closer to working-class radicalism.

During the election, Babbage and his supporters used the *Economy of Machinery* as a manifesto. The supporters viewed his scientific reputation as a crucial asset, but in the event its effects were mixed. He had travelled the country looking at manufacturing techniques, but this was countered by shouted jibes at the hustings that he was 'only an amateur'—that he did not know manufacturing from the inside, as a practitioner.[52] His condemnation of the publishing trade was especially offensive. Conservative *Blackwood's* complained that 'Booby Babbage' 'sneers at the booksellers, as a body of men, in his late dirty duodecimo', and was shocked that a Tory government were 'such confounded asses' as to give him 7,000 pounds for his 'toy', the notorious calculating machine.[53] The story of the machine made him easy to brand as an ingenious but impractical projector, who failed to recognize the generous support governments had provided for his research.

Many readers were uncomfortable with the praise *Decline* heaped on foreign governments that gave honours, money, and status to scientific men. Was the situation on the Continent, where naturalists and mathematicians served as ministers of state, really to be envied? Was Alexander von Humboldt's greatest triumph his scientific research in Latin America, or his service as chamberlain to the King of Prussia? At best, such posts could be seen as financially beneficial distractions, and at worst they were menial and degrading. As William Wordsworth complained to the mathematician William Rowan Hamilton, if Babbage was right, it appeared that the hill of science could only be ascended by state functionaries with 'pockets laden with cash'.[54] Davy's successor at the Royal Institution, Michael Faraday, sponsored publication of a pamphlet written by 'A Foreigner' (the Dutch physicist Gerard Moll), which pointed out that the mathematician Pierre-Simon Laplace, after serving unsuccessfully as Minister of the Interior under Napoleon, had been appointed by him as Chancellor to the French Senate, where the only requirement was 'the necessary abnegation to put his signature indiscriminately under any decree, however unjust or oppressive'.[55] Additional honours might not do any particular harm, but most men of learning cared little about medals and ribbons. Such 'toys', it was pointed out, were of less importance for scientific men than for the governments that awarded them.

In fact, the only area where Babbage called for more state interference in national affairs was for exceptional treatment for men of science engaged in discovery. This, however, was mocked as special pleading, and critics quickly pointed out the dangers. Who would choose those to receive benefits? How would the recipients negotiate the hazard, all too evident in France, of avoiding interference from their patrons? Science was not a profession in England, but many felt that the models being proposed would make it a 'job'.

For Babbage, though, criticisms about potential jobbery missed the point, for they exhibited a lack of understanding of the depth of the changes advocated in the book. The stress on education involved transformations which would take a long time, at least a generation. As his later books emphasized, the spread of knowledge among 'the masses of mankind' would improve efficiency and prepare the masses for 'the reception of other truths'. The suppression of ignorance and fraud, and the promotion of the use of knowledge for human needs, would best be served by secular—not religious—education; for among the unenlightened, even true religion was tinged by 'superstition' and developed irrational habits of thought.[56] *Decline* had shown how reform, underpinned by the new machine economy, would not only get rid of drudgework and error in production and management. It would also sweep through the institutions of learning, rendering them efficient engines for reshaping the minds of the governing classes on a template of reason, transmitting and encouraging discovery and innovation. Babbage's greatest hopes were for the sons of manufacturers, who would have the requisite wealth, skill, and background to benefit from an education in the principles of science. Ministers of state, with minds shaped by reason rather than tradition, would then be in a position to give scientific men the opportunities they needed for the research that would reshape the economy.

The leading force for change in the emerging machine economy, Babbage believed, would be scientific discoverers—the handful of men engaged in original research who represented the highest level of human intelligence. As a discoverer and inventor himself, Babbage was fascinated by the qualities of his own mind. Later in life he bequeathed his brain to the Royal College of Surgeons. One half is still in their museum today, where it is classified among other brains as an anatomical specimen; the other half is in the computing galleries of the Science Museum, where it sits in a display about calculating machines. In 1852, the renowned travelling lecturer Thomas Leger examined Babbage using a novel philosophical instrument called a 'magnetoscope' to determine the relative strength of the power emanating from the different parts of his brain. Babbage's outstanding qualities as a man of science—his caution, combativeness, self-esteem, and lack of imitation—were apparent because they were manifest as measurable magnetic forces.[57] For his part, Babbage kept his magnetoscope reading as a curiosity, for he had no faith in mental forces or attempts to correlate mental faculties with specific parts of the brain. He traced his interest in how things worked to his earliest childhood (if he did not understand one of his toys, he took it apart), but remained sceptical about the role of heredity in mental formation.

In comparing Davy and Wollaston, *Decline* had stressed that their accomplishments were also the result of educated habits of mind rather than special qualities of eye and hand. This was clear enough in the case of Davy, celebrated for his ambitious speculations; but Babbage argued that it was also true of a great observer like Wollaston, whose achievements rested on 'the admirable training of his intellectual faculties' rather than superhuman bodily senses. Because of this, scientific discovery was not 'a gift of nature to a favoured few' but democratically available to any 'patient inquirer into truth'.[58]

Such truth-seekers would have a critical role in the new economy, for ultimately their work would produce the machines that would successively replace higher and higher levels of intelligence. Through publication of *Decline,* Babbage hoped to achieve the first steps in this immense project, looking to long-term educational reforms to reshape and recalibrate contemporary attitudes towards politics and management. He crafted an idealized relation between author and reader, bound together in the perfect communication of meaning, to show what this future might accomplish. *Decline* may read like an ill-tempered invective, but it is grounded in a utopian sense of the possibilities of the machine economy. Even the humblest labourers would, though education, be carried upwards on the tide of progress.

For Babbage, the gradation of intelligence extended throughout the universe. As explained in the conclusion to the *Economy of Machinery and Manufactures,* of all the works of divine power familiar to the natural philosopher on earth, 'the masterpiece of its skill' was man, whose highest attribute was human reason. But this was only the beginning. On other planets still loftier forms of intelligence will have appeared, the product of the same laws of nature:

> But however large the interval that separates the lowest from the highest of those sentient beings which inhabit our planet, all the results of observation, enlightened by all the reasonings of the philosopher, combine to render it probable that in the vast extent of creation, the proudest attribute of our race is but, perchance, the lowest step in the gradation of intellectual existence. For, since every portion of our own material globe, and every animated being it supports, afford, on more scrutinizing enquiry, more perfect evidence of design, it would indeed be most unphilosophical to believe that those sister spheres should each be no more than a floating chaos of unformed matter;—or, being all the work of the same Almighty architect, that no living eye should be gladdened by their forms of beauty, that no intellectual being should expand its faculties in decyphering their laws.[59]

And what is the end of this utopian vision? Would the need for original discoverers like Babbage, Davy, and Wollaston ever completely disappear? Of course it would: for even God's greatest works—human reason, discovery, and genius—could ultimately be enacted by a process akin to that to be seen in the calculating machine. As Babbage explained to guests when demonstrating a part of his engine at his famous soirées, a long sequence of numbers could appear in order, only to change suddenly and apparently at random: but this was a programmed change foreseen by the designer at the beginning.[60] What looked to ordinary observers like miracles of creation could be understood by the man of science as the intelligent actions of a divine machine.

3

The Conduct of Everyday Life

John Herschel's *Preliminary Discourse on the Study of Natural Philosophy*

> 'It'the littherary wontherr of the wurrld,' says he; 'and sure your lordship must have seen it, the latther numbers ispicially—cheap as durrt, bound in gleezed calico, six shillings a vollum. The illusthrious neems of Walther Scott, Thomas Moore, Doctor Southey, Sir James Mackintosh, Docther Donovan, and meself, are to be found in the list of conthributors. It's the Phaynix of Cyclopajies—a litherary Bacon. . . . fild with the pure end lambent flame of science.'
>
> —William Makepeace Thackeray on the *Cabinet Cyclopaedia* in *Fraser's Magazine*, 1838

In the 'Yellowplush Papers'—the satirical memoirs of a footman—that established his reputation, William Thackeray parodied the prolific science writer Dionysius Lardner, who was extolling the virtues of his *Cabinet Cyclopaedia*. Lardner's long-running series did feature many 'illusthrious' writers, and among the most highly regarded was the astronomer and mathematician John Herschel. His *A Preliminary Discourse on the Study of Natural Philosophy*, which outlined the principles and practices of the physical sciences, had been hailed on its publication early in 1831 as the greatest work of its kind since the writings of the seventeenth-century philosopher and statesman Francis Bacon—hence Thackeray's reference to 'litherary

Bacon'. Bacon's works on natural philosophy had been printed as lavish folios, however, while the *Preliminary Discourse* was distinctly unprepossessing, being 'cheap as durrt' and flimsily 'bound in gleezed calico'.

Thackeray's parody suggests the need to take a closer look at Herschel's famous work as an inexpensive book that was bought, borrowed, and read by tens of thousands of readers. Seen from this perspective, the significance of the *Preliminary Discourse* turns out not to be what might have been expected. A few readers viewed the book as a contribution to philosophy, as the starting point for a personal vocation in science, or as an attempt to create a hierarchical network of scientific observers. But given its low price and large sales, readers of the *Preliminary Discourse* were far more likely to have used it as a conduct manual. During the early decades of the nineteenth century, conduct manuals had become staples of publishers' lists, intended particularly for the commercial classes and those new to metropolitan society. Conduct manuals gave instructions, not only about table etiquette and topics for conversation, but also about good character and appropriate modes of thinking. The *Preliminary Discourse* was seen to encourage behaviour based on an understanding of reason as grounded in the practices of science. It showed how ideas of truth could be understood as inhering in the readers who appreciated the 'advantages and pleasures' of the proper pursuit of science.

Science was pervasively bound up with defining and maintaining canons of behaviour, from cultivating appropriate modes for discussion to encouraging the avoidance of outright fraud. Since the seventeenth century, the qualities of moral probity, truth-telling, and religious faith that had characterized the ideal natural philosopher were also the qualities that had traditionally marked the gentlemen. The *Preliminary Discourse* recast this equation for the rapidly changing circumstances of the early nineteenth century. Spelt out for the first time in an accessible book at an affordable price, these qualities could now provide a foundation for good character across the social spectrum.

A Cyclopaedic Discourse

To understand a work like the *Preliminary Discourse*, it is important not to personify it as 'Herschel', but to look first at its material form. Otherwise the name becomes a shorthand for reconstructing the author's intentions, as expressed in a text abstracted from the physical form of a book. Herschel's intentions have been interpreted in different ways and with great sophistication, but for the most part have resulted in the text being located in a tradition of philosophical reflections on induction and metaphysics. The *Preliminary Discourse* is thus seen as the expression of a set of precepts relating to scientific method, with positions uneasily poised between those of the metropolitan empiricist John Stuart Mill on the one hand and the idealist Master of Trinity College, William Whewell, on the other.[1] As a philosophical work advocating induction from particular facts to general theories, the book is also seen to underwrite a hierarchical vision of social organization within science, in which networks of observers pass their findings to a limited circle of authoritative specialists.

There is, however, a problem with these views. The *Preliminary Discourse*, as Thackeray's mockery indicates, does not look like a book of serious philosophy, and certainly would not have appeared as one to most of those who read it when it first appeared. Books of philosophy were almost always substantial well-produced volumes, with generously spaced lines of large and readable type, following a format developed in the second half of the eighteenth century by Scottish publishers such as Thomas Cadell and William Strahan. The pattern was to publish such treatises first in quartos or folios, suitable for the aristocratic and academic libraries, and then as less-expensive but still substantial and well-produced octavos, available to a somewhat wider range of readers, particularly those in the professions. This pattern of publication, although challenged in many areas, continued to dominate the first issues of serious original works not only in philosophy, but also in history, economics, and the sciences.[2] It was the form in which John Murray issued many of Humphry Davy's writings (though not *Consolations*), and the first two editions of Charles Lyell's *Principles of Geology*. Such works, with their select audiences, were rarely in a position to benefit from economies of scale demanded by newer, smaller, and cheaper formats.

Herschel's book, in contrast, was read by his contemporaries in the context of the changes that were sweeping through the publishing world in the 1820s and early 1830s. As a 'preliminary discourse', it was part of the long tradition of introductory works at the start of major encyclopaedias, best exemplified in Jean le Rond d'Alembert's similarly titled work to the great French encyclopaedia of the eighteenth century. Issued in many volumes and in different formats, such encyclopaedias reached huge audiences, offering surveys of all of human learning. They had flourished in Britain after publication of the *Britannica* from 1771, and especially from the early years of the nineteenth century. Besides the ongoing editions of *Britannica*, there was *Rees's Cyclopaedia* (commenced 1802), the *Edinburgh Encyclopaedia* (from 1808), and the *Encyclopaedia Metropolitana* (started in 1817).

These large projects had featured books in large formats suitable for genteel libraries. The early success of the Society for the Diffusion of Useful Knowledge, however, indicated the existence of a market for encyclopaedic works affordable by middle- and working-class audiences. Longman, by some distance the largest publisher in science, medicine, and related fields, projected an ambitious series of small books, grouped together as a 'Cyclopaedia' and edited by Dionysius Lardner. With generous financial incentives from Longman, Lardner invited authors of great reputation to contribute accounts of their special areas. The first to appear, in 1829, was a two-volume history of

Scotland by Sir Walter Scott. Between then and 1844, a total of 133 volumes appeared, organized into various 'libraries' or 'cabinets'. These included a Historical Series, a Series on Arts and Manufactures, a Scientific Series, a Biographical Series, and a Geographical Series. Each was to have an introductory volume surveying the subject and outlining its methods and special concerns. The prospectus announced the purpose: the work would comprise, in 'a determinate number of volumes', all of human learning, 'a complete library of instruction, amusement, and general reference'.[3]

In mid-January 1830, when asked by Lardner to write the preliminary volume for the series on the sciences, John Herschel was in his mid-thirties and had a high reputation among men of science, not only as the son of the celebrated astronomer William Herschel (discoverer of the planet Uranus, or 'Georgium Sidus') but also in his own right. Top scorer on the mathematics examination at Cambridge in his year, he had published important papers on double stars and various aspects of mathematics. He was, however, not known as a writer for the general public. For example, his articles on 'Light' and 'Sound' for the *Encyclopaedia Metropolitana* were 'introductory' only to specialist students. On a scheme designed by Samuel Taylor Coleridge, lengthy contributions to the *Encyclopaedia Metropolitana* were published separately as convenient octavos as well as appearing as double-columned entries within the large quarto *Encyclopaedia*. But Herschel's *Metropolitana* articles, as some of the most technically demanding in the whole work, were never issued in this separate format. Herschel explored with Lardner the possibility of issuing a more accessible treatise on light for the *Cabinet Cylopaedia*, and in 1829 they had agreed that Herschel would write the treatise on astronomy.[4] To Lardner Herschel appeared as the ideal author for the critically important introductory work. The variety of Herschel's achievements demonstrated not only an ability to survey large and complex fields, but also expertise across an unusual range of scientific subjects. Most contemporary men of science saw themselves as chemists, geologists, astronomers, experimental natural philosophers, or mathematicians: Herschel had expertise in all these fields. Lardner signed him up for 250 pounds, a typical fee for contributors, and insistently pushed his author towards delivery of the text.[5]

In its organization as separate treatises rather than short articles, the concept of the *Cabinet Cyclopaedia* was in many ways similar to that of the *Metropolitana*. The smaller format, however, clearly marked out individual volumes as separate treatises from the beginning (Fig. 9). From the publisher's perspective, these had the signal advantage that individual titles could be sold separately from the start. Books such as Herschel's *Preliminary Discourse* therefore appeared as distinctive works, not dependent upon a sequence of parts which might not coincide with the beginning and end of the work. Longman accordingly advertised the volume for sale in three different ways: as part of the whole *Cyclopaedia*; as the lead work in the series on natural philosophy; and as an independent treatise. Every copy included three title pages so that readers could make their choice. Herschel clearly wanted to put his expertise in the background, insisting that he appear on these title pages simply as 'John Frederick William Herschel, Esq., A. M., Late Fellow of St John's College, Cambridge, &c. &c. &c.', without an 'ostentatious parade' of honours and societies trailing in its wake.[6] (His editor had exposed himself to mockery of reformers like Babbage by appearing as 'the Rev. Dionysius Lardner, LL.D., Professor of Natural Philosophy and Astronomy in the University of London; F.R.S.E., Hon. F.S.P. Camb.; F. Ast. S. L.; Hon. F.S.A. Scot.; M.R.I.A.'.)[7] In appearing as the author of a book for a popular audience, Herschel wanted to make his mark as a gentleman; it was left to the reviewers to stress his expertise, which they did.

A tell-tale sign of the cost-cutting measures involved in producing the volumes of the *Cyclopaedia* is the type. *Consolations in Travel* and *Decline of Science* were printed by Thomas Davison on hand presses

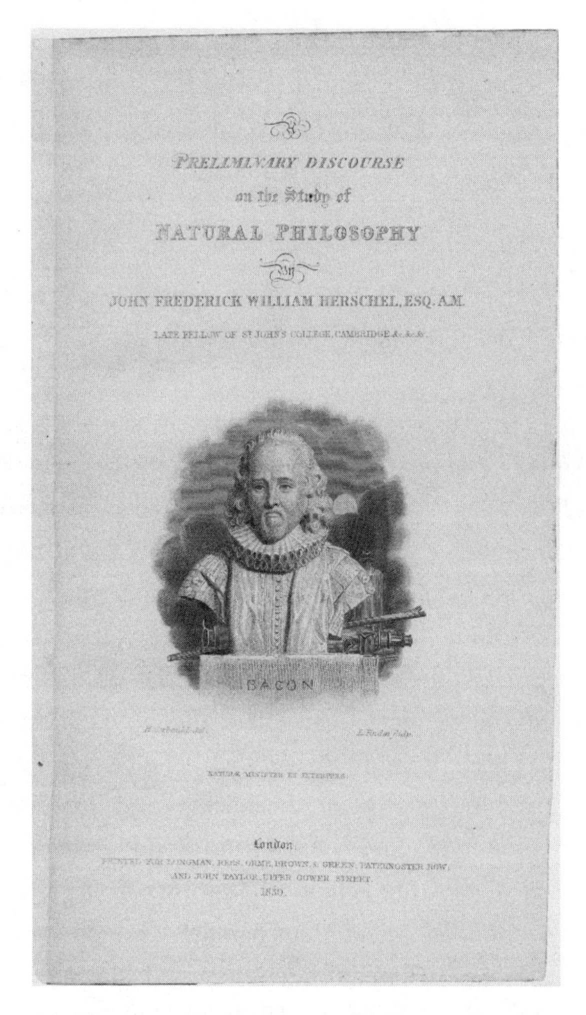

Fig. 9. Illustrated engraved title page from John Herschel, *A Preliminary Discourse on the Study of Natural Philosophy* (London: Longman, 1830 [1831]). Herschel's book had two additional title pages, making it possible for the purchaser to highlight its place in Lardner's *Cyclopaedia,* its place within the scientific series, or its character as an independent work.

using a very fine type and good paper; the *Preliminary Discourse,* printed by Alexander Spottiswoode, was a steam-press production using a cheaper type and lower quality paper, as those who have enlarged it for modern reissues have tended to discover. Some of the loss of definition is probably attributable to stereotyping, which involved making a mould of the forms for the book, an effective way of saving money when a work was likely to go through many printings, because the pages did not need to be freshly set up in type each time. Along with machine printing, stereotyping was one of the innovations that was transforming printing from an artisanal craft to a machine-based industry. Another important indication of the attempt to widen the readership was the binding. The *Preliminary Discourse* is an early example of a case binding, employing a covering of glazed pink calico cloth over boards made of card. Wealthy purchasers or libraries could have it rebound, usually in leather, but those for whom six shillings was already a lot of money had a presentable volume ready for use.

Success in publishing such a work depended entirely upon economies of scale; and by just about any standard, this work was a success. The records of sales are partial, due to destruction of some of the Longman archive during World War II. Nonetheless, we know that the first issue, of over 7,000 copies, sold out in a few months, 2,000 more were printed in March 1831, and 2,000 more in June. The years 1832 and 1833 saw further issues of 2,000, and the imprint dates of further issues show that the work sold well for many years, at least until the middle of the nineteenth century and later than that in the United States. As with *Consolations,* it also was frequently translated, and there were French, German, Italian, Swedish, and Russian editions.

A Manual of Conduct

In the hothouse world of scientific London, the turmoil over reform of the Royal Society reached a crescendo in the spring and summer of 1830. The *Preliminary Discourse* is usually considered separately from this debate, but in fact it emerged directly from it. Herschel had known about Babbage's plans to write a book on the decline of science from the beginning, and supported many of its reforming aims. At the last minute he had added a footnote about the state of science to his article on 'Sound' for the *Encyclopaedia Metropolitana,* contrasting the French *Annales de chimie* and the great scientific reviewing journals in the German lands with the poverty of their equivalents in Britain. 'It is in vain to conceal the melancholy truth,' he wrote, 'We are fast dropping behind. . . . The causes are at once obvious and deep seated.'[8] Babbage asked for a copy of the note in proof, which Herschel sent in March 1830. At this time Herschel also read Babbage's draft manuscript, and his reply was unequivocal: 'If I were near you and could do it without hurting you and thought you would not return it with interest I would give you a good slap in the face.'[9] Rather than sending the manuscript to the press, Babbage should consign it to the flames: 'Burn it, or rewrite it.'

Herschel never joined in the public furore over *Decline,* and in many ways he did not need to, as Babbage did the job for him. The brief comments in the article on 'Sound' were blazoned forth in *Decline*'s preface along with comments from Davy and other luminaries, as testimony that science was in a bad way. Herschel, a quiet man who hated public controversy, deeply regretted mentioning the subject in print. As Herschel complained to Whewell, his words had been wrenched out of the restricted context in which they had appeared, providing 'a text for Babbage to preach upon'. In the ensuing controversy the footnote from the article on 'Sound' was repeatedly reproduced, with a readership far beyond those who would have seen the original. It immediately rendered Herschel (like Davy) a leader of a 'declinist' party. Unhappy about the way his name had been used, Herschel never joined in any of the polemics raised on his behalf. He lamented to the chemist John F. Daniell, 'I could scarcely have supposed at the time of writing that note, that it would ever have acquired the prominence which has been conferred upon it in the course of a controversy in which I have abstained from taking any part:—still less that its object and motive should have been so far misinterpreted.'[10] It was only with the greatest reluctance that he allowed his name to go forward in November 1830 to contest the election of the non-scientific Duke of Sussex as President of the Royal Society. After his motives and manners in challenging the Duke—the King's brother—had been questioned in the newspapers, Herschel's failure to be elected came as a great personal relief.[11] Herschel supported reform, but feared the execution in *Decline* a disaster from which it would be difficult for science to recover any shred of reputation. The resulting controversy, he told the Irish geologist William Fitton, 'by systematically lowering the motives of all scientific men' had required them more than ever to be above suspicion.[12]

As Herschel realized in writing the *Preliminary Discourse*, the issue of conduct was central. If *Decline*, with its polemical tone and political edge, was an example of how not to behave, the *Preliminary Discourse* could offer a model for the actions of the ideal seeker after truth. The most obvious way to achieve this was through a tone, structure, and literary style in every way different from the polemics that had been spewing forth for months in the newspapers. In accepting Lardner's commission for a reflective overview of the sciences in February 1830, Herschel was already committed to establishing the claims of the scientific vocation as a way to truth, and the need to do this became only more imperative after he read *Decline* in manuscript a few weeks later. The circumstances could hardly have been more highly charged, for Herschel wrote most of the work in the summer of 1830, immediately after the clashes over Babbage's book and before the contested election for the presidency in the autumn. Written in the eye of a political storm, *Preliminary Discourse* offered a quietly utopian vision of science and its public uses.

Readers found in the *Preliminary Discourse* an ordered enterprise, in which the different parts of knowledge had clear relations to one another and to the whole. The text is divided into three parts. The first, 'On the General Nature and Advantages of the Study of the Physical Sciences', provides a gateway to the subject, explaining the purpose of a knowledge of nature and the correct frame of mind in which to pursue it. The second part examines the general principles used in the successful practice of the physical sciences, issues of observation, classification, induction, and generalization. The final part summarizes the results that had been achieved in the different arenas of the physical sciences. As the book explains, the structure of the sciences is in continual transformation through discovery at the limits of knowledge. Whenever researchers find a class of phenomena in nature that seems 'elementary' and beyond the range of analysis, the enquiry into those phenomena and their laws becomes 'a separate branch of science'.[13]

The *Preliminary Discourse* is usually considered from philosophical perspectives of observation and experiment, but, as important as these are, it has not been noticed that the entire work is grounded in the assumption that reading has the power to transform the human condition. Readers of the book are told to study nature, and addressed as potential fellow practitioners. But the implication is that such 'study' can occur through the act of reading, and that a scientific understanding can come through the printed page as well as through the eyepiece of a telescope. The narrative of the book is inclusive, giving people the impression that in understanding the *Preliminary Discourse,* they are engaged in the first steps of scientific study. In his address at the opening of the Windsor and Eton Public Library and Reading Room in 1833, Herschel stressed the power of reading for all people, whatever their status or occupation:

> It is hardly possible but the character should take a higher and better tone from the constant habit of associating in thought with a class of thinkers, to say the least of it, above the average of humanity. It is morally impossible but that the manners should take a tinge of good breeding and

> civilization, from having constantly before one's eyes the way in which the best bred and the best informed men have talked and conducted themselves in their intercourse with each other. There is a gentle, but perfectly irresistible coercion in a habit of reading well directed over the whole tenor of a man's character and conduct, which is not the less effectual because it works insensibly, and because it is really the last thing he dreams of.... It civilizes the conduct of men—and *suffers* them not to remain barbarous.[14]

Herschel was a great admirer of Brougham's introductory pamphlet on the *Objects, Advantages, and Pleasures of Science*, as the aims of his own work make clear.[15] The *Preliminary Discourse* is thus not a recipe book for inductive discovery, but an invitation to share, through reading, the scientific frame of mind. The contemplation of general laws—such as those discovered by Isaac Newton—forms 'as it were, a link between ourselves and the best and noblest benefactors of our species'. Through this communion with great scientific minds readers could vicariously engage in the act of discovery and an appreciation of natural truth; and through these also are brought to the frontiers of human knowledge and 'nearer to their Creator'.[16]

These issues are underlined from the beginning of the work, so that the more detailed discussions of optics, heat, astronomy, and so forth will be understood as invitations to active engagement. By starting with the famous epigraph from Cicero, 'Before all other things, man is distinguished by his pursuit and investigation of TRUTH', the *Preliminary Discourse* shows that though man is naked and weak compared to most animals, he is 'the undisputed lord of the creation'. Not only does this place all nature at his disposal, but it also opens up the prospect of pleasures beyond the merely sensual. Man 'contemplates the world, and the objects around him, not with a passive, indifferent gaze, as a set of phenomena in which he has no further interest than as they affect his immediate situation, and can be rendered subservient to his comfort, but as a system disposed with order and design'. Understanding these high ambitions of the pursuit of science allows the reader to realize how little is known, or can be known, while offering protection from unthinking enthusiasm by 'a habit of strict investigation'.[17]

The main purpose of the *Preliminary Discourse* is to develop this 'habit' in its readers. The central section, which discussed 'the rules by which a systematic examination of nature should be conducted', begins by distinguishing necessary truths of mathematics from the contingent ones of 'nature and its laws'. The properties of an ellipse or a triangle are constant and defined; the effects of dissolving sugar in water had to be experienced through the senses, through observation and experiment. A person locked in a dark cell might conceivably reconstruct all of mathematics, but would never be able to reconstruct the content of the different physical sciences.

The danger is that our judgment can be misled, so that the senses can be tricksters or magicians: 'Cross the first two fingers of one hand, and place a pea in the fork between them, moving and rolling it about on a table, we shall (especially if we close our eyes) be fully persuaded we have two peas.'[18] Put the right hand in ice-water, the left in hot water, and then put them both in water at body temperature: the right hand will be hot, the left cold. Or hold your nose and try to taste the difference between a piece of cinnamon and a shaving of pine wood. Ventriloquists, it is explained, depended for their success on the limits of our sense of hearing. To avoid being misled by these sensory pitfalls, the *Preliminary Discourse* recommends that we be constantly aware of our sensory limitations. Previous opinions need not be abandoned, but they cannot be considered as dogmas, and we must be willing to consider instances to the contrary. Those engaged in science cannot expect to discover ultimate causes—the domain of theology—but we can look for proximate causes, those which lead immediately to the phenomena. These are causes that explain particular phenomena, and are both sufficient and observable in action. Drawing on Isaac Newton's discussion in the *Principia Mathematica* of

1687, the *Preliminary Discourse* identifies these as *vera causae*, literally 'true causes'. The concept of a *vera causa* had been central to the philosophy of 'common sense' as developed by Thomas Reid and others in Scotland during the eighteenth century. Reid's 'common sense' is grounded in our direct experience of observable reality, to be perceived through the senses and reasoned upon according to the basic principles of mind present in all people as an element of their natural constitution. Experience, in the form of carefully observed instances, is critical. The *Preliminary Discourse* gives a striking example from a book of South American travels. A local guide in the Pampas points into the air, and shouts 'a lion!' The traveller, with difficulty, can see condors circling in the distance. The guide knows, through long experience, that there is a dead animal below, being eaten by a lion, while the condors wait their turn to get to the carcass. 'The sight of the birds was to him what the sight of the lion alone could have been to the traveller, a full assurance of its existence.'[19]

In cases like this, the knowledgeable local guide provides an accessible model of how inferences can be made. It is the kind of knowledge that anyone can appreciate, and by associating the hands-on experience of an uneducated non-European, it becomes all the more accessible. In establishing the higher generalizations of science, we are similarly not 'left wholly without a guide', for the 'nature of abstract or general reasoning points out in a great measure the course we must pursue'.[20] What is required is a consideration of the ways that phenomena differ or agree, so that shared qualities stand out as the basis for a general proposition. The reader of the *Preliminary Discourse*, however, is not encouraged to develop a philosophically consistent theory of induction, but to appreciate the 'kind of mental convention' that makes entering into the practical work of science possible.[21]

In mature sciences, juxtaposition and comparison can be carried out 'where facts are numerous, well observed, and methodically arranged' by numerous workers through a division of labour. In developing sciences, in contrast, the facts are less familiar, requiring a single person who unites wide knowledge with 'individual penetration'.[22] Even the casual reader, with no appreciation of mathematics, was in a good position to understand the basic findings of astronomy or mechanics, although discovery was only open to those who possessed the requisite skills in reasoning. In forming theories, we are not 'abandoned to the unrestrained exercise of the imagination, or at liberty to lay down arbitrary principles', but rather the same rules of conduct that governed our earliest inductions equally apply to the highest. Our responsibility is to act not as slaves 'broke loose from fetters', but as freemen who have 'learned the lessons of self-restraint in the school of just subordination'.[23]

The example of how modesty should operate in science is drawn from printing:

> A sheet of blank paper is placed upon a frame, and shoved forwards, and after winding its way successively over and under half a dozen rollers, and performing many other strange evolutions, comes out printed on both sides.... We might frequent printing-houses, and form a theory of printing, and having worked our way up to the point where the mechanical action commenced (the boiler of the steam-engine), and verified it by taking to pieces, and putting together again, the train of wheels and the presses, and by sound theoretical examination of all the transfers of motion from one part to another; we should, at length, pronounce our theory good, and declare that we understood printing thoroughly.[24]

Readers could imagine themselves extending the enquiry further, to understand the origin of the power that ran the piston rod of the engine. Here the answers were less certain. Perhaps the source was 'the den of some powerful unknown animal'? The need for water and fuel, 'the breathing noises', and the sheer power generated, pointed in this direction. Or perhaps another theorist, glimpsing the fire and hearing the sound of boiling, might think the cause was steam, and 'form a theory more in consonance with fact'. What neither theorist could be expected to discover soon, the *Preliminary Discourse* cautioned, was why boiling water produces this kind of 'active energy'. Such a cause was precisely the sort likely long to remain a secret, and for the time being effectively beyond science.[25]

The printing press also served to demonstrate lessons about how individual discoveries could push forward the development of science. Although new instruments such as the telescope were critical, the *Preliminary Discourse* stressed that the invention of the printing press had ensured the progress of science since the Reformation. Glimmers of understanding had appeared before this point, so that 'The moment this took place . . . the sparks of information . . . began to accumulate into a genial glow, and the flame was at length kindled which was speedily to acquire the strength and rapid spread of a conflagration.' Scientific journals allowed discoveries to spread from one place to another, and become part of a permanent record. A particularly significant role was played by introductory works—implicitly, volumes such as those of the kind the reader was consulting from the *Cabinet Cyclopaedia*—which brought together recent discoveries under a common point of view, and suggested future directions for research.[26]

Most readers of the *Preliminary Discourse* were unlikely to make theories about printing presses, steam engines, or the principles of heat. What the book offered was a wider rational foundation for how to behave in everyday life.

> There is something in the contemplation of general laws which powerfully persuades us to merge individual feeling, and to commit ourselves unreservedly to their disposal; while the observation of the calm, energetic regularity of nature, the immense scale of her operations, and the certainty with which her ends are attained, tends, irresistibly, to tranquillize and reassure the mind, and render it less accessible to repining, selfish, and turbulent emotions. And this it does, not by debasing our nature into weak compliances and abject submission to circumstances, but by filling us, as from an inward spring, with a sense of nobleness and power which enables us to rise superior to them . . .[27]

These words draw on a tradition of the scientific pastoral, in which the contemplation of nature leads to an inner repose and the erasure of selfish, individual feelings. It is a projection of a scientific ideal, in which knowledge is above what Herschel's associate, the Cambridge geologist Adam Sedgwick, decried as 'the dreary wild of politics'.[28]

At the same time, the escape from the passions afforded by reasoned contemplation ought to revitalize political life, enabling collective action unimpeded by greed and individual short-sightedness:

> for why should we despair that the reason which has enabled us to subdue all nature to our purposes, should (if permitted and assisted by the providence of God) achieve a far more difficult conquest; and ultimately find some means of enabling the collective wisdom of mankind to bear down those obstacles which individual short-sightedness, selfishness, and passion, oppose to all improvements, and by which the highest hopes are continually blighted, and the fairest prospects marred.[29]

The reader, in appreciating the results and methods of science, is drawn on by the highest human ambition to realize the limits of knowledge and the bounds of human ability. Capable of understanding only some of nature's work, the reader is invited to contemplate

> a Power and Intelligence to which he may well apply the term infinite, since he not only sees no actual limit to the instances in which they are manifested, but finds, on the contrary, that the farther he enquires, and the wider his sphere of observation extends, they continually open upon him in increasing abundance; and that as the study of one prepares him to understand and appreciate another, refinement follows on refinement, wonder on wonder, till his faculties become bewildered in admiration, and his intellect falls back on itself in utter hopelessness of arriving at an end.[30]

Fired by 'an unbounded spirit of enquiry', those pursuing science ultimately will appreciate the need for a higher power, and avoid the brazen certainty of unbelief.

Elegant Extracts and Philosophical Readers

These transcendent aspects of the physical sciences, especially astronomy, were already familiar to readers, but their meaning could be ambiguous. Talk of the stars was considered to be one of the most effective ways to attract the opposite sex.[31] It was equally well established that turning one's eyes towards the sky was a useful device for snubbing old friends and acquaintances, as the print, 'The Cut Celestial' from 1827 (Fig. 10) demonstrates. Like the London dandy, depicted in the latest fashion, viewers of the print who wished to 'cut' a 'dunning Tailor, or story-telling Uncle, or a Respectable Man with a Shabby-drest Wife and poodle-dog' were instructed: 'You are suddenly struck with the beauty of the Heavens! What a magnificent Structure Herschell-Georgium Sidus! By that time your Tailor is gone by—and you pursue your walk—solus.'

At the opposite end of the social spectrum, the contemplation of philosophy could be depicted as a distraction from the everyday necessities of eating, sleeping, and work. In a best-selling print entitled 'Frontispiece for the Penny Magazine', Charles Grant depicted a series of vignettes (Fig. 11) that showed readers pursing knowledge to the neglect of their ordinary occupations. Sailors reading navigation are drowning, gardeners reading botany are trampling on their hot-frames, sweeps reading divinity are sitting atop their chimneys. 'Philosophy' shows an unshaven, ill-clothed man lounging outside an eating house, having spent his supper money on the *Penny Magazine* to learn what men like Herschel have to say. The apothecary's boy reading physics instead of delivering 'physic' is about to demonstrate the law of gravity by falling through an open cellar.

The association of science with lofty distraction could easily be parodied, but it also provided an invaluable resource for books which aimed to integrate science more thoroughly with ideals of good character. And in this the *Preliminary Discourse* was extremely successful.

Fig. 10. A 'dandy' of the 1820s, unable to pay his bills, coolly observes the heavens rather than acknowledging his deferential tailor. As the accompanying text noted, 'Cut the first—is the Celestial: When you meet your dunning Tailor, or story-telling Uncle . . . You are suddenly struck with the beauty of the Heavens! What a magnificent Structure Herschell-Georgium Sidus: By that time your Tailor is gone by—& and you pursue your walk—solus.' M. Egerton, 'The Cut Celestial', aquatint (London: Thomas M'Lean, 1827).

Fig. 11. Parodies of the reading of 'useful knowledge' often highlighted the effects of choosing books and periodicals over less intellectual fare. 'Philosophy' and 'Physics' from Charles Jameson Grant, 'Frontispiece to the Penny Magazine of the Society for the Diffusion of Useful Knowledge. Vol 1' (London: E. Lacey, 1832).

Reviewers familiar with Herschel's previous writings, such as the writer in the religious monthly *Wesleyan-Methodist Magazine,* were confirmed in their opinion of the author's character and accomplishments; many were surprised given that his articles for the *Metropolitana* had been so abstrusely mathematical. The highest praise was that he was the transparent voice of nature: 'nothing, in short, that indicates a profound determination to speak to the initiated only, or to restrict the exterior limits of popular information, by precluding the multitude from listening to the common voice of wisdom, and understanding all the noble lessons which she teaches'. The convenient format and accessibility of the book led the *Monthly Magazine* to praise it as 'the most exciting volume of the kind we ever met with'. The style was admired as 'clear', 'vigorous', 'manly'.[32]

It is not surprising that the book was regularly mined for quotations. Newspaper editors often looked to the *Preliminary Discourse* for odd or unusual facts to fill up newspaper columns, but the prefatory discussions were the most frequently repeated. These were extracted in the reviews—ranging from weeklies such as the *Literary Gazette*, the *Examiner*, and *Athenaeum* to long quotations in the *Quarterly* and *Edinburgh*—and remained a popular resource for editors of annuals and collections of 'elegant extracts'. This serial anthologizing pointed out 'beauties' and literary 'gems', processing texts in a way that allowed long books to be integrated into the conversation. Herschel, like most authors, knew that this would happen. At a personal level, he was already all too familiar with the process, demonstrated so clearly by what had happened to the declinist footnote in his *Metropolitana* article on 'Sound'. More generally, however, he was aware that this process of extracting informed the writing of any book intended for multiple and wide audiences.[33] Readers, as the *Quarterly Review* pointed out, would 'find scattered through the work instances of vivid and happy illustration, where the fancy is usefully called into action, so as sometimes to remind us of the splendid pictures which crowd upon us in the style of Bacon, where the restlessness of the intellect appears to keep its chamber-fellow, the imagination, perpetually awake'.[34] Not all the reviewers appreciated Herschel's attempt at a bold style. The generally favourable notice in the conservative *Fraser's Magazine*, which admitted a suspicion of all writing for 'popular libraries', found that these passages 'sound a little hollow and complimentary; and seem more to resemble laboured declamation than the free and spontaneous utterance of real conviction'.[35]

Nonetheless, such passages were widely reproduced, becoming set-pieces to be learned by heart. The positioning of such key passages in the book was vital to establishing suitable patterns of contemplation in a wide range of readers, both women and men. 'It seems to have been his great aim,' *The Times* noted of Herschel, 'to render every part of his "discourse" intelligible to the general reader, and to bring down philosophy from that region of clouds and mists in which it is often concealed from the eyes of ordinary men to the every-day scenes of life.'[36] Herschel's success in achieving this aim is evident in the way that quotations from the book were routinely incorporated into cheap weeklies peddling useful knowledge for a penny. Both the *Penny Magazine* and its religiously orthodox rival the *Saturday Magazine* often featured pithy paragraphs from the work.

This process of clipping, cutting, and extracting is equally marked in learned works, for quoting the *Preliminary Discourse* became a way of associating a text with some of the qualities of a classic. The book became so well known that extracts were reproduced without attribution, or identified only as being by 'a profound philosopher' or 'an elegant writer'. The most significant examples are in writings by Charles Darwin, particularly in *On the Origin of Species* and the *Journal of Researches . . . during the Voyage of the HMS* Beagle.[37] The language of these works, especially when contemplating the purpose and larger meanings of science, is pervaded by references to the *Preliminary Discourse.* The most celebrated instance is on the first page of the *Origin,* which refers to 'that mystery of mysteries, as it has been called by one of our greatest philosophers'.[38]

Darwin's case as a reader is telling here in a wider sense. There can be little doubt that he hoped his own scientific work—especially on the controversial subject of species—would attract Herschel's attention, if not his outright personal approval. The *Preliminary Discourse* established Herschel as something of a philosophical arbiter, a judge of probity in dealing with theory and speculation, because it defined and validated science as a suitable vocation for aspiring young men like Darwin. Darwin could not (and did not) learn much about induction from the second part of the *Preliminary Discourse* that he did not already know through direct experience of practising natural history at Cambridge, Edinburgh, and at home in Shrewsbury. But he could absorb lessons about the nature of the scientific life. For Darwin, as much as for the readers of the penny weekly press, Herschel's book was not a rulebook of procedures, but an inspiration to find his own scientific vocation. As he later recalled at the end of his life, it 'stirred up in me a burning zeal to add even the most humble contribution to the noble structure of Natural Science'.[39]

It was in February 1831 that the young Darwin—completing his residence requirement at Christ's College in Cambridge—discovered the book, then just published a fortnight earlier. 'If you have not read Herschel in Lardners Cyclo,' he told his cousin William Darwin Fox, '—read it directly.'[40] Darwin's father had been concerned that his youngest son would end up as a genteel dilettante, unable to focus on a single object or to make anything out of his life. The *Preliminary Discourse* gave Darwin confidence that his interest in science was a high calling. The historian Sir James Mackintosh, a close family friend of the Darwins and Wedgwoods, affirmed to those in his circle that no other man in Europe than Herschel could have written the book. In a footnote to his own contribution to the *Cabinet Cyclopaedia*, Mackintosh endorsed Herschel's book as 'the finest work of philosophical genius our age has seen', a remark that was endlessly recycled in advertisements on both sides of the Atlantic.[41]

Clearly the *Preliminary Discourse* was 'philosophical', but to read the book primarily as a treatise on the question of induction obscures the larger context, both of the text itself and the settings in which it was used. In its framing and address to the reader, it is clear that the book is a much more complex work, aimed at cultivating appropriate actions and patterns of thought, rather than detailing systematic doctrines. It belongs very much in the tradition of works by Francis Bacon and Robert Boyle, seventeenth-century authors celebrated for presenting science as involving both the character of the observer as well as the means of enquiry. The format was ideal for the purposes of moral guidance. Other than Bacon's *Novum Organum* of 1620, the most important model for the *Preliminary Discourse,* and one cited frequently in its notes, was musician, artist, and author William Jackson's *The Four Ages* of 1798, which employed the universal history of mankind as a framework for presenting pithy thoughts on questions of morality and character.[42] The use of history was closely based upon Jackson's example, which tied the progress of knowledge to the wider fortunes of the civilizing process.

If the *Preliminary Discourse* was not originally a work of philosophy in our sense, how did it become one? The answer to this question is obvious once we look at how the book was read by those who considered themselves as authors in an explicitly philosophical tradition, and who at this time were projecting the possibility of tackling systematically the special problems posed by science. The journalist, economist, and philosophical author John Stuart Mill, writing anonymously in the *Examiner,* praised the *Preliminary Discourse* as the 'least incomplete view' of philosophical truth and the means of discovering it, that had 'yet been produced'.[43] As this rather back-handed compliment suggests, the author was being placed very precisely as a man of science. Depicted by Mill as being free from the limits of vision of most scientific enquirers, who often dismissed philosophizing as 'frivolous and idle', Herschel was at the same time clearly of their number, and would need to press his ideas far further and more systematically to produce a philosophical work. Mill was here advertising his own ongoing philosophical project, which he began writing soon after publishing his notice in the *Examiner,* but he was also drawing a line between the practice of philosophy and the practice of science. This became even clearer in 1838 when Mill's *System of Logic, Ratiocinative and Inductive* lauded the *Preliminary Discourse* for its 'admirably-selected' examples of induction in the physical sciences, but criticized its philosophy as insufficiently developed.[44]

From a very different philosophical and political standpoint, William Whewell made the same point. His review of the *Preliminary Discourse* in the *Quarterly Review* praised the book as 'the first attempt since Bacon to deliver a connected body of rules of philosophizing which shall apply alike to the conduct of all researches directed to the discovery of laws of nature' and 'a most valuable addition to our philosophical literature'. At another level, though, the review sharply distinguished the role of the philosopher from that of the man of science. The distinction Whewell drew between the philosopher and the man of science was between two types of mind, one formulating overarching rules, the other carrying out 'the detail of this labour'. These activities involve different skills, different vocations, and different habits of mind. 'The two characters differ,' Whewell argued, 'like that of a great general and of a good engineer.'[45] From this perspective, Herschel clearly belonged in the latter category, for he was 'principally known by abstruse and technical writings, unapproachable by the common reader'. This gave his 'less professional views' in the *Preliminary Discourse* a special interest, but it did not make its author a genuine philosopher. The same could be said, the review notes, of the French expert on fossil bones Georges Cuvier and the Swedish chemist Jacob Berzelius, both known for technical writings, but also for overviews—masterly but not truly philosophical—of the progress of science. In contrast, Whewell placed himself in the ranks of philosophers, and his review of the *Preliminary Discourse* in the *Quarterly* was the first substantial outline of philosophical views he went on to elaborate during the next few years in five fat volumes.

Philosophers, Scientists, and Barbarians

This distinction between philosophy and science is at the heart of the invention of the word 'scientist'. At the meeting in Cambridge of the British Association for the Advancement of Science in 1833, the poet and metaphysician Samuel Taylor Coleridge had complained that men of science, concerned with practical issues and experiments, were unjustly usurping a higher calling in identifying themselves as 'natural philosophers'. This was a claim very much along the lines of his general critique of the march of intellect and the cultural authority increasingly accorded to utilitarian disciplines such as chemistry and geology. A new word was needed, and Whewell, drawing an analogy with 'artist', responded at the meeting by suggesting 'scientist'. The term 'scientist' was invented not because of the increasing authority for the role, but as a way of maintaining the higher and prior claims of philosophy, which Whewell (like Coleridge) saw as his true vocation. It did not help matters that in publicizing his new coinage Whewell seemed unable to resist making further analogies with 'journalist', 'sciolist', 'atheist', and 'tobacconist', roles scarcely to be emulated.[46] The ironic intent behind the comparison with 'sciolist', meaning a superficial, pretentious attitude towards scholarship, would not have been lost on contemporaries.

Given its character as a putdown, it is unsurprising that 'scientist' did not catch on. Herschel never used the term, Babbage ignored it, and practitioners from Michael Faraday in the 1830s to the mathematical biologist D'Arcy Wentworth Thompson in the 1920s continued calling themselves 'natural philosophers' and 'men of science'. The geologist Sedgwick scribbled against a passage in a book by Whewell claiming such a word was needed, 'better die of this want than bestialise our tongue by such barbarisms'.[47] Half a century later the professional comparative anatomist Thomas Henry Huxley noted that for 'any one who respects the English language, I think "scientist" must be about as pleasing a word as "electrocution"'.[48] The barbarism did become widely used in the United States from the 1840s, where those engaged in practical work in laboratories and observatories were less troubled by philosophical aspirations. Only several decades later did 'scientist' travel back across the Atlantic to be used in Britain, and it remained controversial well into the twentieth century.[49]

Whewell quickly acknowledged that his suggestion 'was not generally palatable'. The *Preliminary Discourse,* considered as a conduct manual, suggests why this was the case. Herschel's focus on induction showed how philosophy grew out of and was part of science. That is why Herschel saw himself as a natural philosopher, and why his friend Whewell saw him as a 'scientist', akin to a 'great engineer'. From Whewell's perspective, the reflections on method in a book like the *Preliminary Discourse* were too closely bound up with daily life, useful invention, and practical outcomes. One of his few criticisms of Herschel's book was that it focused too much on utility. These characteristics, as Whewell reflected in the months after the British Association meeting at which he coined the word 'scientist', were peculiarly prominent in the works of male authors. 'Notwithstanding all the dreams of theorists,' he wrote, 'there is a sex in minds.' Whewell thought the circumstances were rare, but when a woman did write from deep knowledge, she would do so not as a scientist, but with the lucid clarity and single-mindedness of a philosopher.[50]

4

Mathematics for the Million?

Mary Somerville's *On the Connexion of the Physical Sciences*

> This is in the very spirit of that compression, by which an octavo volume of mathematics is brought within the compass of a three-penny pamphlet, and, at the same time, simplified from the intellectual standard of the well-read student in physics to the mind of a mechanic . . .
>
> —Anonymous review of Mary Somerville, *Mechanism of the Heavens, Athenaeum* (21 January 1832)

In its weekly issue for 29 March 1834, the *Mechanics' Magazine*—a self-help journal of invention and discovery—announced a new book that it believed merited a place on the shelf alongside Herschel's *Preliminary Discourse.* Written by the mathematician Mary Somerville, *On the Connexion of the Physical Sciences* drew together a range of subjects that were undergoing unprecedented change. It revealed the common bonds between the sciences at a time when they were being carved up into distinct disciplines. Reviewers praised *Connexion* for bringing together the sciences in a new way which avoided the materialism for which the French had been condemned. It was not a textbook—when first written, no one taught this range of subjects in a university or school course—but rather an outstanding example of the reflective treatises characteristic of the years around 1830. *Connexion* was original, not in announcing discoveries, but in reviewing a subject in a novel and compelling way. In successive editions, it encouraged attempts to bring British experimental approaches and continental mathematical analysis together. It played an important role in debates about theories of light and experimental studies of electricity and magnetism. Through its wide readership, it became a key work in transforming the 'natural philosophy' of the seventeenth and eighteenth centuries into the 'physics' of the nineteenth, demonstrating the unity of the observable world. In fact, the book was so important that the *Mechanics' Magazine* said it should not be on a bookshelf at all: 'Instead of that we say—read it! read it!'[1]

In the aftermath of Hans Christian Oersted's sensational discovery of a relation between electricity and magnetism in 1820, there were tantalizing indications of a unity underlying all physical phenomena. Other previously separate powers, such as gravity, light, and atomic attraction might also be knitted together. But connecting the sciences was not easy. At a practical level, the emergence of sharper boundaries between disciplines meant that practitioners often failed to talk with one another, with researches being published in separate journals, in distinctive institutions, using incompatible methods and increasingly esoteric languages. *Connexion* intervened in this complex and rapidly developing situation. As the great natural philosopher James Clerk Maxwell later recognized, *Connexion* was among those 'suggestive books, which put into a definite, intelligible, and communicable form, the guiding ideas that are already working in the minds of men of science, so as to lead them to discoveries, but which they cannot yet shape into a definite statement'.[2]

Mathematics was presented in *Connexion* as the most promising source of ultimate unity, but, then as now, it was often thought to be inaccessible to all but a few. Moreover, experimental phenomena relating to electricity and magnetism seemed difficult to fit into the mathematical model that governed the Newtonian system, which relied on central forces and the dynamical laws governing bodies in motion. Readers of *Connexion* thus faced a paradox: the book expressed a vision of science based in mathematics, but without using a single equation. Clearly a non-mathematical synthesis of all physical science was feasible only in part, but how was it possible at all? For contemporary readers, the problem was made even more prominent by the fact that the author was a woman.

Mathematics as Useful Knowledge

The origin of *Connexion* can be traced back to the ambitious schemes of the Society for the Diffusion of Useful Knowledge to bring science to the people. As part of its plan, the Society had projected an extensive 'Library of Useful Knowledge', a series of inexpensive works in thirty-two-page parts at sixpence each, that summed up the findings of scientific and technological fields. These were published by Baldwin & Cradock of Paternoster Row in London, on inexpensive paper and usually in closely packed double columns. The series had been introduced by the politician and man of science Henry Brougham's preliminary discourse on the *Objects, Advantages, and Pleasures of Science,* which outlined the virtues of scientific investigation, as 'the most dignified and happy of human occupations' for men and women. This had been released to great acclaim and huge sales in March 1827.[3]

At this propitious moment for popular education, seven years before *Connexion* was published, the Society announced its intention to commission an accessible version of the greatest scientific monument of the age, Pierre-Simon Laplace's five-volume *Mécanique céleste,* published in French from 1798 to 1827. This was a mathematical synthesis of the most developed and 'perfect' of all the sciences, celestial mechanics. Isaac Newton's *Principia Mathematica* of 1687 had shown that the unseen attractive force of gravity, operating according to divine providence, governed both the heavens and the earth, from the movements of Saturn to the dropping of a stone. The *Mécanique céleste* took this much further, showing that no outside intervention was required to keep the Newtonian system in motion. All the movements of the earth, moon, sun, and planets could be derived from a single equation for the motion of matter. Anyone who wished to read this work, however, needed to have mastered the highest levels of analytical calculus developed in France during the eighteenth century. The final volume of Laplace's great work attempted to extend this approach, explaining sound, molecular attraction, and other phenomena, through a system of short-range forces. Although more speculative, this was a mathematical vision of what the future of the sciences might hold. Laplace, who had died early in March 1827, was seen to have a superhuman status, 'rising over the all the Great Teachers of mankind'.[4]

For the projectors of the Society for the Diffusion of Useful Knowledge, particularly Brougham, making the findings of Laplace's mathematical synthesis available to the people was not only possible, but essential. Two treatises, Brougham suggested, would cover the basics. One would be introductory, no more than a hundred pages, for readers without mathematics; the other would be for more advanced readers. Space was not a problem—the tightly packed pages of the Library of Useful Knowledge could make available the equivalent of 800 pages of ordinary printed matter within the bounds of a single treatise. The critical issue was audience: the aim was to reach 'the unlearned', to 'carry ignorant readers into an understanding of the depths of science'.[5] 'The kind of thing wanted,' Brougham explained, 'is such a description of that divine work as will both explain to the unlearned the sort of thing it is—the plan, the vast merit, the wonderful truths unfolded or methodized—and the calculus by which all this is accomplished, and will also give a somewhat deeper insight to the uninitiated.' However, finding someone qualified to grapple with the *Mécanique céleste* was difficult. 'In England there are not now twenty people who know this great work, except by name,' Brougham noted, 'and not a hundred who know it even by name. My firm belief is that Mrs. Somerville could add two ciphers to each of those figures.'[6] 'I fear you will think me very daring for the design I have formed against Mrs. Somerville,' Brougham wrote to her husband William in March 1827, 'and still more for making you my advocate with her: through whom I have every hope of prevailing.'[7]

Behind this straightforward plan was an almost unlimited hope for the possibilities of readers achieving enlightenment through self-education, even in the most complex and difficult fields of human learning. Early parts of the Library of Useful Knowledge thus included sophisticated discussions of mechanics, hydrostatics, hydraulics, and pneumatics. These required mastery not only of geometry and algebra, but also of calculus, which was taught in Britain at this time exclusively at the universities of Cambridge and Edinburgh. Brougham, who had published two papers in the Royal Society's *Philosophical Transactions* before he was nineteen, was himself an accomplished mathematician. He believed that an understanding of proportion, number, and relation had been implanted in the human mind, so that just as bees could make perfect hexagons according the laws of geometry, so too could ordinary men and women master the principles, step by step, of the full range of mathematics. As Brougham had written in his pamphlet on education, 'It is not necessary that all who are taught, or even any large proportion, should go beyond the rudiments; but whoever feels within himself a desire and an aptitude to proceed further, will press forward; and the chances of discovery, both in the arts and in science itself, will be thus indefinitely multiplied.'[8] On this calculation, it was perfectly possible that thousands of ordinary readers in Britain could, through their own reading and private study, master the higher calculus of the *Mécanique céleste*.

In approaching Mary Somerville, Brougham had in mind not only one of the few in Britain who was mathematically qualified to write such a work, but also someone who exemplified in her person the virtues of self-help that would be required more widely in the population if the proposed work was to succeed. Born in provincial Scotland in 1780 as Mary Fairfax, she combined traditional female accomplishments with a passion for abstract mathematics. Although her mother feared she would become mad or a bluestocking, she was also praised as intelligent and eager to excel. In the eighteenth century both men and women participated in a lively culture of mathematical puzzle-solving in magazines such as the *Lady's Diary* and *Gentleman's Diary*. Having learned of the existence of 'algebra' by seeing a problem of this kind in a fashion magazine, in 1811 Somerville went on to publish elegant solutions in the *New Series of the Mathematical Repository* to other problems under the pseudonym of 'A Lady', her studies guided by leading mathematicians in Edinburgh, including the celebrated author in astronomy and geology John Playfair.[9] Her first husband had not encouraged her studies, but after his death, her second, the physician and traveller William Somerville, was unusually supportive. As a wedding gift, he encouraged the purchase of a selection of advanced (and very expensive) mathematical books. 'I was thirty-three years of age when I bought this excellent little library,' she recalled at the end of her life, 'I could hardly believe that I possessed such a treasure when I looked back upon the day that I first saw the mysterious word "Algebra".'[10]

At the time Brougham made his proposal, the Somervilles were well established in London. They lived in a comfortable house in Chelsea, where William held a lucrative position as physician to the Royal Hospital. The pursuit of knowledge offered a vital niche in elite urban society: the couple collected minerals, made observations, discussed new books, and welcomed visiting foreigners. Among their friends, Alexander and Jane Marcet—the latter known for her *Conversations on Chemistry* of 1806—were key figures, as were other couples such as the geologist Roderick Murchison and his wife Charlotte, and established natural philosophers such as William Wollaston and Henry Warburton. 'All kinds of scientific subjects were discussed,' Mary Somerville recalled, 'experiments tried and astronomical observations made in a little garden in front of the house.'[11] Women played a central role in this sociable milieu. A physician visiting from the United States, Charles Caldwell, made the mistake of claiming that American women were better conversationalists than those in England. A few days later he received a teasing invitation card to an unusual 'conversation party of ladies' at the Somervilles, with a half-dozen guests including the prison reformer Elizabeth Fry and the novelist Maria Edgeworth. Other than William Somerville, Caldwell discovered he was the only man present, and was, Mary Somerville told him, 'soon about to have female conversation enough to convince you that the ladies of England can talk as well as those of your own country'.[12]

In their engagement with science, the Somervilles were not aspiring amateurs, nor potential patrons (they did not have the money for that); rather, they were valued participants in a common enterprise. Thus, even though Mary Somerville's contributions to the *Mathematical Repository* were published, this was not their primary significance; instead, they indicated participation in a network of practice that depended only partly on print.[13] Other scientific activities grew out of the same sociable setting. In the long summer days of 1825, Somerville carried out delicate experiments to show a connection between magnetism and sunlight, by exposing a sewing needle to the violet range of the spectrum. Like many experiments in English science at this time, these relied upon simple materials readily to hand or borrowed from friends, rather than an expensive laboratory. The results were exciting. They were communicated to the Royal

Society of London by her husband William, who also managed her other public affairs. The paper was read in February 1826, and published in the *Philosophical Transactions* soon afterwards. All contributions to the Royal Society were handled in this way, without discussion or questions, and Somerville had no wish to give a public performance, even had she been allowed into the Royal Society. To publish under her own name was therefore a big step. Publication, however, not only made the results available to a much wider range of readers, but also vindicated science as an informal, genteel activity pursued by a wide range of practitioners. Somerville and her reforming Whig friends were aware that she could contribute to science, not least as an example of what a woman could accomplish. As her friend Captain Henry Kater wrote to Oersted, 'such a woman must be considered as an honor to any country'.[14] Somerville presented herself in ways that made her iconic role particularly suitable. In private, she could have a sharp wit and a fine sense of the ridiculous. Having grown up opposing almost everything her parents believed in, among close friends she was forthright about her unorthodox religious and political views and enthusiastic in discussing the latest scientific findings. With those not in the inner circle, however, she held back, maintaining a facade of quiet modesty.

Brougham approached William Somerville about the Laplace project because to ask a married lady directly would have been improper. Mary Somerville shared Brougham's belief in the possibilities of education, though she suspected that a single work, or even two, could never accomplish the task. But she agreed to try. She would assume knowledge of calculus, physical mechanics, and astronomy, but would add explanations and later results, clarifying much that had been left obscure. The lucid formulations and elegant proofs, so evident in her problem solutions for the *Mathematical Repository*, would be brought to bear on the greatest scientific work of the age.

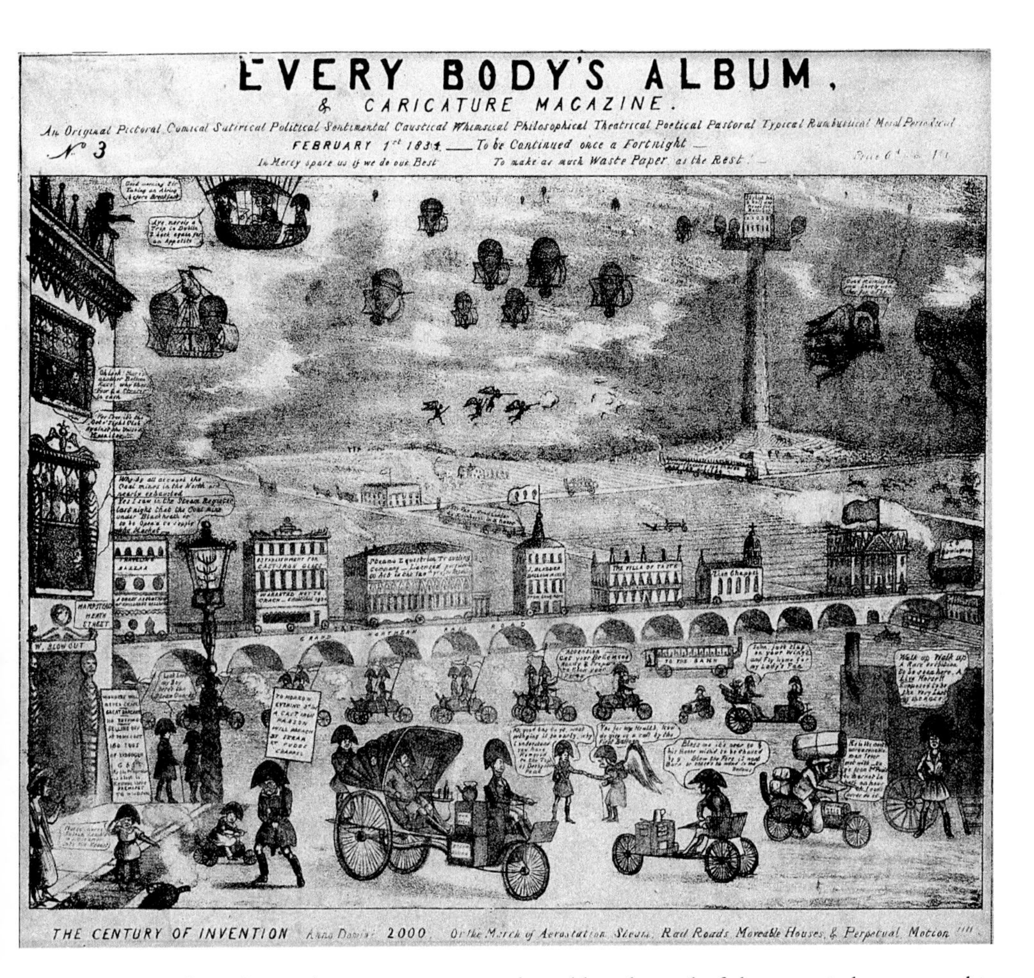

PLATE 1 Looking forward to a steam-powered world at the end of the twentieth century, this lithograph appeared in one of the most successful caricature magazines of the early 1830s. Charles Jameson Grant, 'The Century of Invention Anno Domini 2000, Or the March of Aerostation, Steam, Rail Roads, Moveable Houses, & Perpetual Motion', *Every Body's Album, & Caricature Magazine*, no. 3, 1 Feb. 1834.

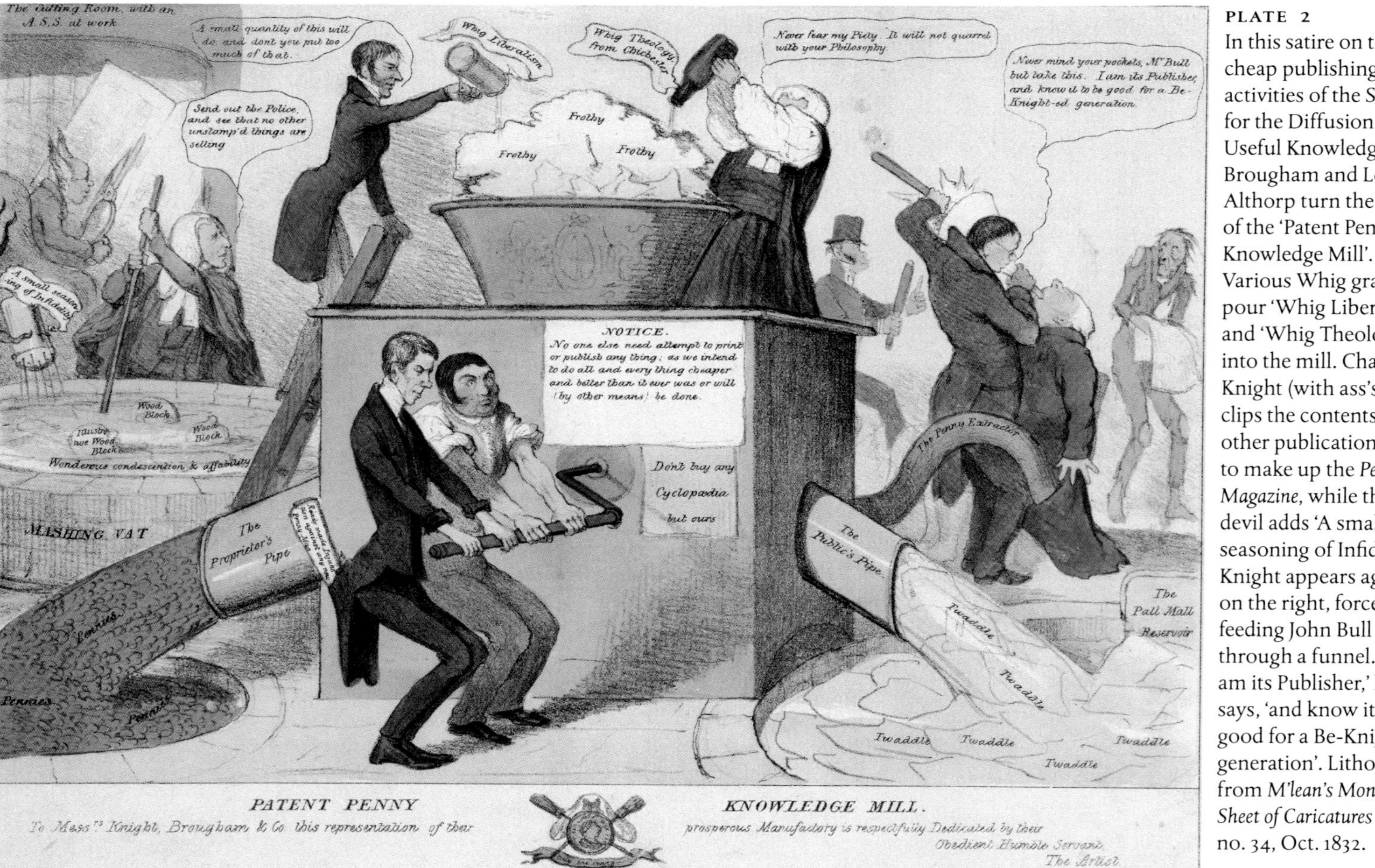

PLATE 2
In this satire on the cheap publishing activities of the Society for the Diffusion of Useful Knowledge, Brougham and Lord Althorp turn the crank of the 'Patent Penny Knowledge Mill'. Various Whig grandees pour 'Whig Liberalism' and 'Whig Theology' into the mill. Charles Knight (with ass's ears) clips the contents of other publications to make up the *Penny Magazine*, while the devil adds 'A small seasoning of Infidelity'. Knight appears again on the right, force-feeding John Bull through a funnel. 'I am its Publisher,' he says, 'and know it to be good for a Be-Knighted generation'. Lithograph from *M'lean's Monthly Sheet of Caricatures* no. 34, Oct. 1832.

PLATE 3
This street scene from London's poorest district anticipates the chaos that will be produced by technical advances and universal education. While street musicians play, a common labourer carries a classical figure on his shoulder and readers (both wealthy and poor) use the street as an impromptu reading room. A broken-down cart advertises 'Calais in five hours' and a placard offers trips to York by 'steam coach'. New buildings under construction, covered by advertisements for useful knowledge, block the view of church spires. 'Shortshanks' (Robert Seymour), 'The March of Intellect', etching (London: Thomas M'Lean, *c.* 1828).

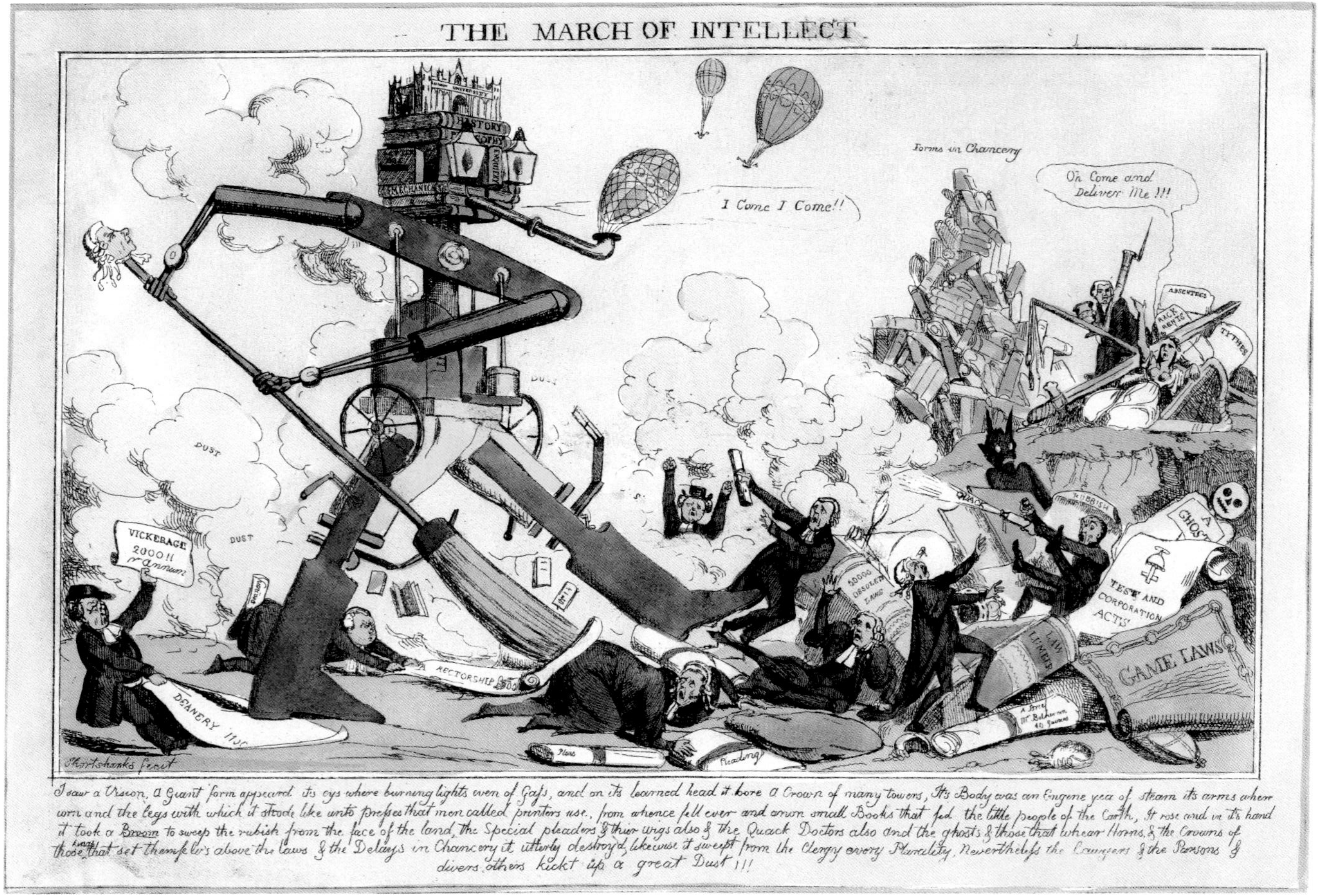

PLATE 4
'I saw a Vision, a great form . . . 'The giant engine of intellect, crowned by the recently built University of London and spouting balloons through its pipe, sweeps away sinecured placemen and antiquated laws with a 'Brougham'. 'Shortshanks' (Robert Seymour), 'The March of Intellect', etching, *c.* 1828.

PLATE 5
The 1820s were noted for the emergence of clearly defined 'types', here identified as the 'Sentimental', the 'Narcissical', the 'Byronical', and the 'Ironical'. The final comment is left to the 'Ironical': 'My Parrot shall talk Sentiment, my Monkey make Love, & my Poodle write Poetry when all our intentions are completed.' 'Shortshanks' (Robert Seymour), 'Modern Style of Language. Sketches from a Fashionable Conversazione', etching (London: Thomas M'Lean, 1828).

PLATE 6

A lecturer on phrenology addresses a dubiously fashionable audience. Without a wig, his head is revealed as covered in bumps. Busts and books surround him, most of them punning various aspects of the new science. The images on the wall illustrate various puns ('life's a bumper'); particular phrenological faculties ('abstraction', 'suspicion', and 'prying'), as well as the availability of 'patent phrenological hats'. Henry Thomas Alken, 'Calves' Heads and Brains: Or a Phrenological Lecture', etching, 1826.

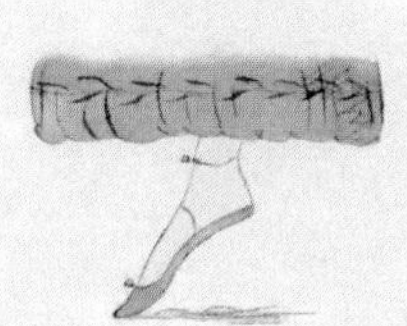
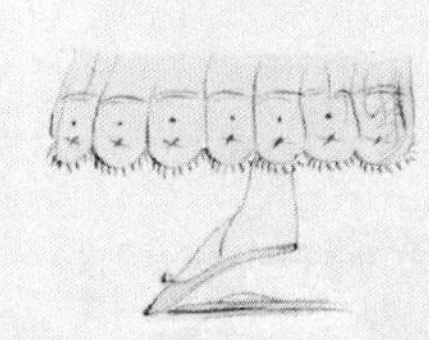
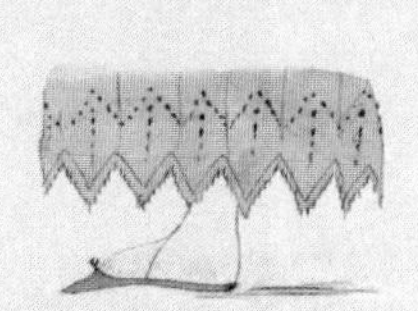
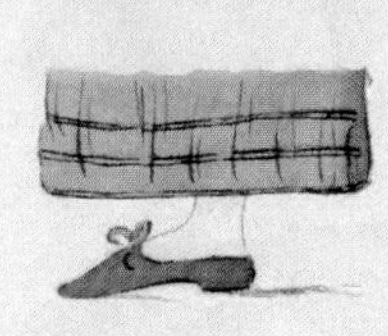

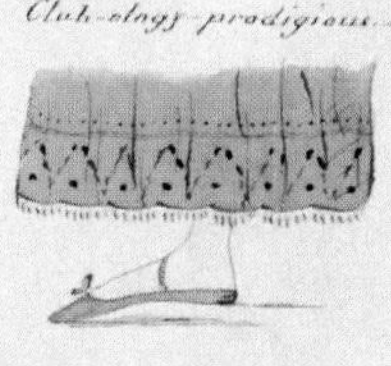
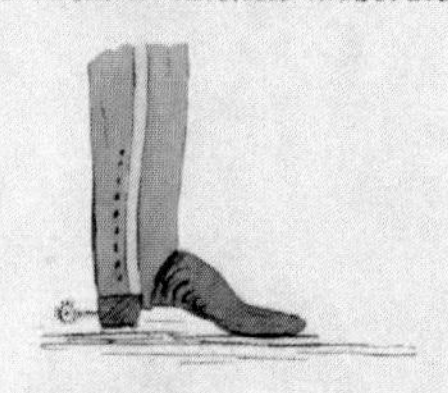
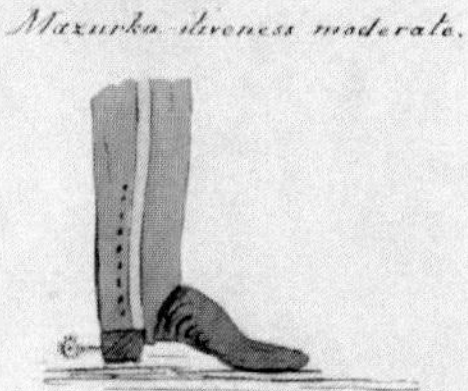

PLATE 7 The fashionable science of footwear: from 'Gossip-pativeness very full' to 'Combativeness very large.' [Catherine Sinclair], 'Toe-tology—versus—Phrenology.—A Course of Lectures to be Delivered (at the "Foot" of Arthur's Seat.) on the 1st of April, to Prove that There is More Expression in the Foot, than in the Head.—', Edinburgh, Lazars, *c.* 1830.

(a)

(b)

PLATE 8
At the time 'Sartor Resartus' appeared in *Fraser's*, the cloth trades were at the leading edge of political activity and debates about trade unionism. (a) The organized tailor, or 'Flint', is shown here with aggressively open scissors and a pointedly phallic bodkin. His head is a cabbage, symbol of tailors' rights. (b) The second illustration offers a tamer image, with arms crossed and the scissors safely shut; the effect is enhanced by the delicacy of the lithography and the fineness of the colouring. 'A Flint', engraving (London: Thomas Tegg, 1811) and 'The Tailor', coloured lithograph (London: G. E. Madeley, *c.* 1830).

For the first two years Somerville wrote her treatise for the Library of Useful Knowledge in secret, planning to destroy the work unless competent authorities (first Brougham and then John Herschel) approved of what she had done. By the end, though, she was characteristically robust. When Herschel conveyed in March 1830 the potentially devastating news that an American astronomer had just published the first volume of a complete annotated English translation of Laplace's work, Somerville was unfazed and stressed the difference between what she was doing and a straight translation: 'I rather wish to state principles clearly, and to arrive at the results by as easy methods as possible, than to enter into all the mathematical details. I daresay you think me very bold, but I do feel inclined to proceed and to get it into the press as soon as soon as possible.'[15]

Somerville decided to deal only with the first two volumes of the *Mécanique céleste,* but in other respects the manuscript she presented to Brougham matched the agreed commission, including the length, level of treatment, and division into two parts. A seventy-page 'Preliminary Dissertation', written by Somerville herself, provided a concise non-mathematical overview; the main body of the work, on the other hand, was an uncompromising account of what Laplace had done, in 610 pages of dense mathematics.

The book, however, was not to be published under the auspices of the Society for the Diffusion of Useful Knowledge. Nor was the promised abstract of Newton's *Principia.* Brougham had overestimated the public appetite for abstruse science in inexpensive form, and sales of the Library of Useful Knowledge had fallen off catastrophically. His *Advantages, and Pleasures of Science* had sold 31,200 copies by the end of 1829, but the more specialized volumes were doing far less well, frequently not covering their costs.[16] Several years later the Society for the Diffusion of Useful Knowledge would get around to producing an introduction to the differential and integral calculus in twenty-five parts, but this was clearly intended for university courses, even though Augustus De Morgan, the author, suspected that 'the child of an artisan' or 'a savage' would be better placed to understand the work than middle- and upper-class students.[17] Laplace's mathematics were another order of difficulty, however, and the main body of the manuscript Brougham had in his hands showed that no matter how clear Somerville's exposition, the mathematics was far too complex and abstract for the projected audience. At about this time, the radical quarterly *Westminster Review* predicted that the plan to bring the *Mécanique céleste* before the general reading public, let alone self-educated working people, would fail:

> If the great work of Laplace were placed in the hands of one ignorant of mathematics, he might be able to read the words it contains, and to give the proper names to the various technical signs employed, but not understanding one sentence, he would go through no intellectual labour—his understanding would be unemployed to the same extent as it would be, were he to read a book in a language of which he comprehended not a word. . . . The ignorant must be taught by examples, not by generalities. In other words, an analytic not a synthetic method should be pursued.[18]

Although the *Westminster* was writing in the context of an ongoing attack on the limits of Brougham's support for political reform, even the most optimistic supporters of the diffusion of knowledge were realizing that the value of Somerville's labours would not be in making French celestial mechanics accessible to the millions, but as a symbol. To demonstrate the openness of science to a wide range of practitioners and the virtues of self-help it was of some significance that Somerville's name appear on the title page. However, existing works in the Library of Useful Knowledge were anonymous, which in this case would have obscured the extraordinary and politically significant fact of the book's authorship by a self-educated woman. Somerville was already an icon of reform when Brougham had made his invitation—celebrated throughout Europe for her mathematical accomplishments. She shared his belief that the diffusion and encouragement of learning could assist the higher classes in reasserting their authority to govern the nation. Older forms of aristocratic conduct, such as the heavy gambling and sexual freedom enjoyed by the Prince Regent and Georgiana, Duchess of Devonshire, were increasingly viewed with disfavour. As standards of morality changed and the movement for social reform grew, a woman who combined virtuous domesticity with knowledge of Greek, Latin, and advanced mathematics represented a valuable political asset.

Brougham's unexpected change of heart about publishing Somerville's book produced a crisis, for without Society for the Diffusion of Useful Knowledge backing, such an abstruse mathematical work might never see the light of day. Fortunately, the publisher John Murray, a friend of the Somervilles, agreed to proceed, and after a complex process using the specialist mathematical compositors at the printer William Clowes, publication of the *Mechanism of the Heavens* took place in the final months of 1831. The book's physical appearance was very different from what it would have been as a serial work in the Library of Useful Knowledge (Fig. 12). Rather than being issued inexpensively in parts, with double columns of text and cheap wrappers, the *Mechanism of the Heavens* was beautifully printed in the form traditional for scholarly mathematical books, in a linen cloth casing. The price was one pound and ten shillings, probably three times what it would have cost if issued by the Society for the Diffusion of Useful Knowledge. Brougham estimated that Murray would sell 1,500 copies, but Murray himself more cautiously called for a run of 750. The audience for mathematics was limited, and initial sales were even worse than expected. Three years later, however, the Somervilles were pleasantly surprised to learn that the whole edition had sold out. Encouraged, Mary Somerville completed the manuscript of a second volume, completing the discussion of Laplace's work, but Murray turned it down. Even at Cambridge, the leading mathematical centre in the country, only the most advanced students were

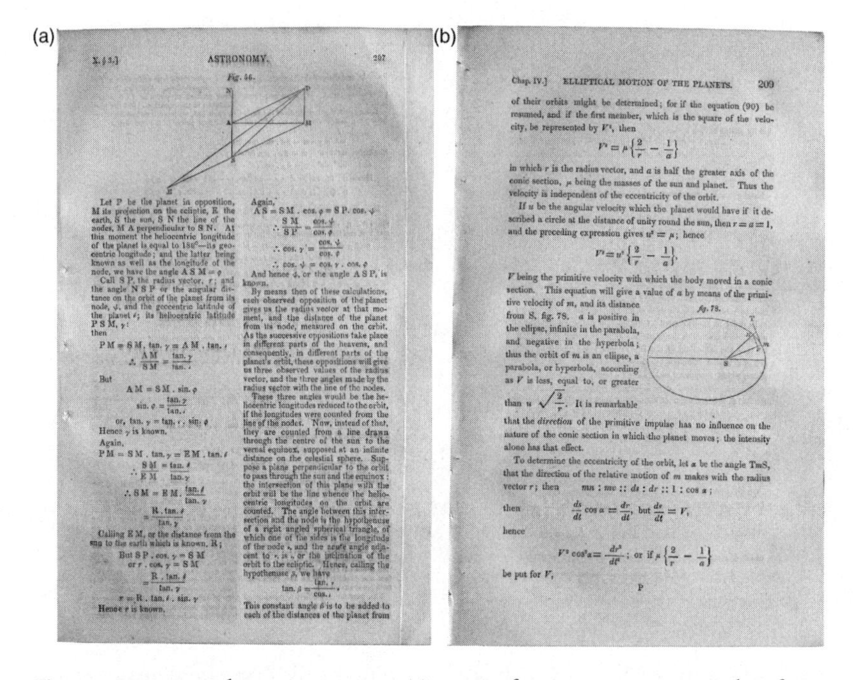

X. § 3.] ASTRONOMY. 207

Fig. 56.

Let P be the planet in opposition, M its projection on the ecliptic, E the earth, S the sun, S N the line of the nodes, M A perpendicular to S N. At this moment the heliocentric longitude of the planet is equal to 180°—its geocentric longitude; and the latter being known as well as the longitude of the node, we have the angle A S M = ϕ

Call S P, the radius vector, r; and the angle N S P or the angular distance on the orbit of the planet from its node, ψ, and the geocentric latitude of the planet ϵ; its heliocentric latitude P S M, γ:

then

$$PM = SM . \tan. \gamma = AM . \tan. \epsilon$$

$$\therefore \frac{AM}{SM} = \frac{\tan. \gamma}{\tan. \epsilon}$$

But

$$AM = SM . \sin. \phi$$

$$\sin. \phi = \frac{\tan. \gamma}{\tan. \epsilon}$$

$$\text{or, } \tan. \gamma = \tan. \epsilon . \sin. \phi$$

Hence γ is known.

Again,

$$PM = SM . \tan. \gamma = EM . \tan. \epsilon$$

$$\therefore \frac{SM}{EM} = \frac{\tan. \epsilon}{\tan. \gamma}$$

$$\therefore SM = EM . \frac{\tan. \epsilon}{\tan. \gamma}$$

$$= \frac{R . \tan. \epsilon}{\tan. \gamma}$$

Calling E M, or the distance from the sun to the earth which is known, R;

$$\text{But } SP . \cos. \gamma = SM$$

$$\text{or } r . \cos. \gamma = SM$$

$$= \frac{R . \tan. \epsilon}{\tan. \gamma}$$

$$r = R . \tan. \epsilon . \sin. \gamma$$

Hence r is known.

Again,

$$AS = SM . \cos. \phi = SP . \cos. \psi$$

$$\therefore \frac{SM}{SP} = \frac{\cos. \psi}{\cos. \phi}$$

$$\therefore \cos. \gamma = \frac{\cos. \psi}{\cos. \phi}$$

$$\therefore \cos. \psi = \cos. \gamma . \cos. \phi$$

And hence ψ, or the angle A S P, is known.

By means then of these calculations, each observed opposition of the planet gives us the radius vector at that moment, and the distance of the planet from its node, measured on the orbit. As the successive oppositions take place in different parts of the heavens, and consequently, in different parts of the planet's orbit, these oppositions will give us three observed values of the radius vector, and the three angles made by the radius vector with the line of the nodes.

These three angles would be the heliocentric longitudes reduced to the orbit, if the longitudes were counted from the line of the nodes. Now, instead of that, they are counted from a line drawn through the centre of the sun to the vernal equinox, supposed at an infinite distance on the celestial sphere. Suppose a plane perpendicular to the orbit to pass through the sun and the equinox: the intersection of this plane with the orbit will be the line whence the heliocentric longitudes on the orbit are counted. The angle between this intersection and the node is the hypothenuse of a right angled spherical triangle, of which one of the sides is the longitude of the node ν, and the acute angle adjacent to ν, is ι, or the inclination of the orbit to the ecliptic. Hence, calling the hypothenuse β, we have

$$\tan. \beta = \frac{\tan. \nu}{\cos. \iota}.$$

This constant angle β is to be added to each of the distances of the planet from

Chap. IV.] ELLIPTICAL MOTION OF THE PLANETS. 209

of their orbits might be determined; for if the equation (90) be resumed, and if the first member, which is the square of the velocity, be represented by V^2, then

$$V^2 = \mu \left\{ \frac{2}{r} - \frac{1}{a} \right\}$$

in which r is the radius vector, and a is half the greater axis of the conic section, μ being the masses of the sun and planet. Thus the velocity is independent of the eccentricity of the orbit.

If u be the angular velocity which the planet would have if it described a circle at the distance of unity round the sun, then $r = a = 1$, and the preceding expression gives $u^2 = \mu$; hence

$$V^2 = u^2 \left\{ \frac{2}{r} - \frac{1}{a} \right\},$$

V being the primitive velocity with which the body moved in a conic section. This equation will give a value of a by means of the primitive velocity of m, and its distance from S, fig. 78. a is positive in the ellipse, infinite in the parabola, and negative in the hyperbola; thus the orbit of m is an ellipse, a parabola, or hyperbola, according as V is less, equal to, or greater than $u \sqrt{\frac{2}{r}}$. It is remarkable that the *direction* of the primitive impulse has no influence on the nature of the conic section in which the planet moves; the intensity alone has that effect.

fig. 78.

To determine the eccentricity of the orbit, let α be the angle TmS, that the direction of the relative motion of m makes with the radius vector r; then $mn : mv :: ds : dr :: 1 : \cos \alpha$;

then

$$\frac{ds}{dt} \cos \alpha = \frac{dr}{dt}, \text{ but } \frac{ds}{dt} = V,$$

hence

$$V^2 \cos^2 \alpha = \frac{dr^2}{dt^2}; \text{ or if } \mu \left\{ \frac{2}{r} - \frac{1}{a} \right\}$$

be put for V,

P

Fig. 12. Setting mathematics in type: (a) a page from a six-penny number from the Library of Useful Knowledge (*Astronomy*, published in 1834) contrasted with a page from (b) Mary Somerville's *Mechanism of the Heavens* (1831). The comparison gives some idea of what her redaction of Pierre-Simon Laplace's *Mécanique céleste* would have looked like had it been published in the cheaper, double-columned pages sponsored by the Society of Useful Knowledge. The treatise on astronomy required only knowledge of geometry and trigonometry, while the *Mechanism of the Heavens* demanded an understanding of advanced calculus.

encouraged to consult the *Mechanism of the Heavens.*[19] Utopian hopes for the mathematical enlightenment of the people had passed.

The Common Bond of Analysis

If *Mechanism of the Heavens* showed that mathematics failed to provide the basis for an accessible synthesis, the 'Preliminary Dissertation' with which Somerville had prefaced the work did suggest a way forward. She had had fifty copies privately printed with a new title page, which she circulated to friends who were not in a position to appreciate a gift of hundreds of pages of technical mathematics. Although the 'Preliminary Dissertation' never appeared for sale in Britain as a separate publication (it was pirated in the United States), it was warmly welcomed by friends and correspondents lacking mathematical experience but with a serious interest in the latest scientific developments. As the Scottish dramatist and poet Joanna Baillie remarked, Somerville had 'done more to remove the light estimation in which the capacity of women is too often held, than all that has been accomplished by the whole sisterhood of poetic damsels and novel-writing authors'.[20]

Even before Somerville had completed the *Mechanism of the Heavens,* she thought the 'Preliminary Dissertation' might serve as the basis for a new book. Her initial plan was to enlarge the discussions on astronomy and dynamics, which had been widely praised as a concise non-mathematical introduction accessible to suitably prepared readers. By March 1832, Somerville's husband characterized it to his sister as 'considerably enlarged, mid-popular & I hope it will have a good sale'.[21] Somerville, however, evidently decided to expand the work further and produce a more substantial work that would also cover electricity, light, sound, and other aspects of the experimental sciences. She finished the new sections during a ten-month stay in Paris. Although there was talk of submitting it to Longman for inclusion in Lardner's *Cyclopaedia,* by the time she returned to London in the late spring of 1833 with the completed manuscript of *On the Connexion of the Physical Sciences,* Murray had accepted it for his list.[22]

In February of the following year *Connexion* was advertised at seven shillings and sixpence. The first edition of 2,000 copies quickly sold out, and the book remained in print for over forty years, with ten editions selling 17,500 copies in all. It was translated into German, French, and Italian, and publishers in Philadelphia and New York in the United States issued pirated editions. Later editions

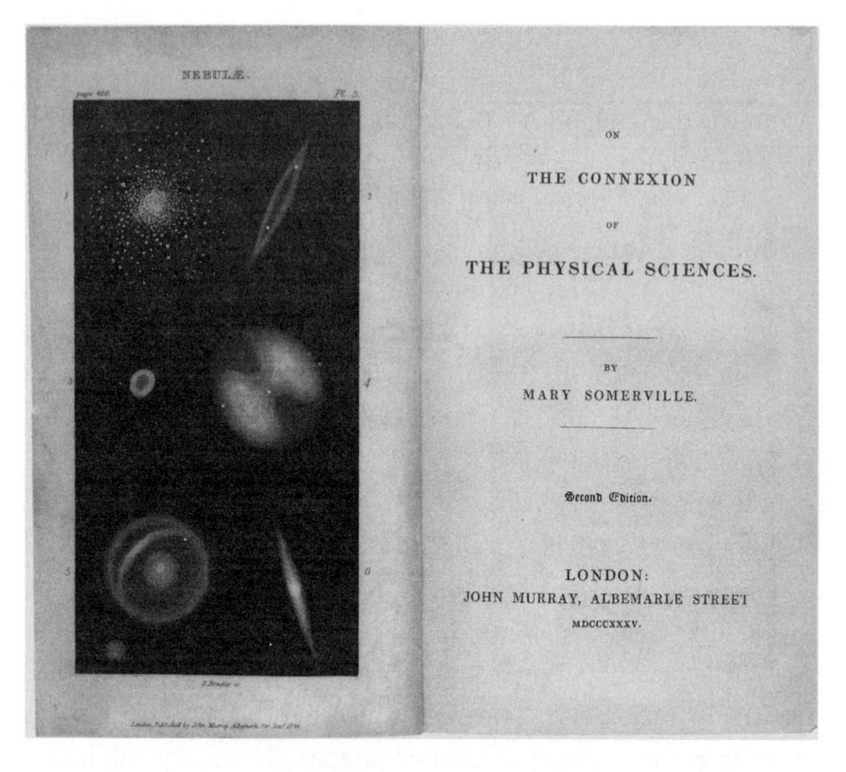

ON

THE CONNEXION

OF

THE PHYSICAL SCIENCES.

BY

MARY SOMERVILLE.

Second Edition.

LONDON:
JOHN MURRAY, ALBEMARLE STREET
MDCCCXXXV.

Fig. 13. This frontispiece illustration of nebulae, added to the second edition of *Connexion*, was part of the publisher's efforts to make the work less abstract and more visually appealing. Mary Somerville, *On the Connexion of the Physical Sciences*, 2nd ed. (London: John Murray, 1835).

of *Connexion* were heavily revised. Although the structure was unchanged, in almost every other way the book was transformed, from its opening sentence to its advocacy of particular theories. Besides notes and illustrations added for clarity, a whole sheet (sixteen pages) had to be added at a late stage to the second edition to report recent experiments on the effects of radiant heat (Fig. 13). As a result of these changes, the later editions became longer, more authoritative, didactic, and ponderous. *Connexion* secured Somerville's reputation among a broad international readership, and copies made their way to schools in India and colonial outposts in New South Wales and the Cape. The illustrated weekly *Saturday Magazine*, like many other British and American periodicals and newspapers, quoted numerous extracts from the book, much in the way they did from *Consolations* and the *Preliminary Discourse*.

The small octavo format, delicate type, and generous line spacing of *Connexion* were typical of other reflective works on science published by Murray during this period, along the model of Marcet's *Conversations on Chemistry* and the cheap editions of Charles Lyell's pioneering *Principles of Geology*. Like Longman in publishing Lardner's *Cyclopedia*, Murray wished to counteract the Society for the Diffusion of Useful Knowledge's inexpensive 'libraries' of Useful Knowledge and Entertaining Knowledge. The chief monument of Murray's efforts was the Family Library, fifty-three volumes of original non-fiction issued between 1829 and 1834. The aim of the Family Library was counter-revolutionary; as one historian has written, it embodied 'a remarkable effort to publish across class lines at a time when class divisions were newly felt to be threatening the fabric of national life'.[23]

Connexion similarly addressed a wide range of readers, from gentlemanly specialists in science to working people eager to tackle something more advanced than an elementary work. The book, as the *Athenaeum* stressed, was 'at the same time a fit companion for the philosopher in his study, and for the literary lady in her boudoir'.[24] Like many of Murray's most successful authors, Somerville was motivated by a reforming zeal that her publisher, as a moderate Tory, did not share. She had long held liberal political and religious views, and believed in governance by a highly educated aristocracy. Her husband campaigned to end corruption at the Royal Society and to elect Herschel as president. Somerville aimed especially to encourage a female readership for her work, part of her lifelong advocacy of the rights of women. The dedication of *Connexion* to Queen Adelaide demonstrates that Somerville aimed to 'make the laws by which the material world is governed more familiar to my countrywomen'. Even as a young girl she had had little patience for the pious attitudes towards female education expressed in the writings of Hannah More and other evangelicals. 'I detested their books,' she later recalled, 'for they imposed such restraints and duties that they seemed to have been written to please men.'[25]

The book moves readers through astronomy, experimental physics, and chemistry in thirty-seven sections. The first fourteen of these, on celestial, lunar, and planetary astronomy, closely follow the text of the 'Preliminary Dissertation' of the *Mechanism of the Heavens*; thirteen further sections then deal with molecular attraction, light, sound, and heat, all topics dealt with in the final volume of the *Mécanique céleste*; there is also some discussion of heat in relation to climate and the distribution of plants, animals, and humans. Sections twenty-eight to thirty-five then survey new developments in electricity and magnetism. The book is rounded off by a discussion of comets, nebulae, and meteorites, with a conclusion on the general laws of matter. These, as also shown in Herschel's *Preliminary Discourse*, were increasingly seen as the contents of the physical sciences.[26]

What, then, was the 'connexion' that held this diverse material together? A brief 'preface' explained the aim of the book:

> The progress of modern science, especially within the last five years, has been remarkable for a tendency to simplify the laws of nature, and to unite detached branches by general principles. In some cases identity has been proved where there appeared to be nothing in common, as in the electric and magnetic influences; in others, as that of light and heat, such analogies have been pointed out as to justify the expectation, that they will ultimately be referred to the same agent: and in all there exists such a bond of union, that proficiency cannot be attained in any one without a knowledge of others.[27]

Despite this promise, readers looking for a single unifying principle or force were disappointed, for the work does not present any totalizing explanation. Instead the book stresses that nature is ultimately unified through a scientific understanding, each part of which linked to the rest.

In the later nineteenth century, with the rise of energy conservation, evolution, and field theory, the kind of links made in *Connexion* lost much of their meaning, and the science writer Arabella Buckley's revision of the work in 1877 was to be the last. A few years before it appeared, the physicist James Clerk Maxwell summed up the problem for readers of his generation: 'when we examine her book in order to find out the nature of the connection of the physical sciences, we are at first tempted to suppose that it is due to the art of the bookbinder, who has bound into one volume such a quantity of information about each of them'.[28]

Somerville, unlike Brougham in his original commission, had no illusions about the difficulties most readers would face in going beyond the basics of geometry and algebra. She explicitly noted in *Connexion* that 'a complete acquaintance' with physics could only be gained by those 'who are well versed in the higher branches of mathematical and mechanical science'. At the same time, though, readers are invited to consider a future in which mathematical knowledge would be a commonplace of education for women and men alike. A passage derived from the 'Preliminary Dissertation' to the *Mechanism of the Heavens*, reassured readers that mastery is

> within the reach of many who shrink from the task, appalled by difficulties, which, perhaps, are not more formidable than those incident to the study of the elements of every branch of knowledge; and who possibly overrate them from disregarding the distinction between the degree of mathematical acquirement necessary for making discoveries, and that which is requisite for understanding what others have done. That the study of mathematics, and their application to astronomy, are full of interest, will be allowed by all who have devoted their time and attention to these pursuits; and they only can estimate the delight of arriving at the truths they disclose, whether it be in the discovery of a world or of a new property of numbers.[29]

Universal mathematical literacy, however, was a future prospect only, so an alternative way of communicating the unity of the sciences was required. The result, in *Connexion* as in the 'Preliminary Dissertation', was a radical innovation in prose. A way had to found to communicate the spirit of mathematical reasoning without formulae or diagrams. All readers, both women and men, were encouraged not just to skim the surface of knowledge, but also to engage in the pleasures of its pursuit.

The demands on the reader, despite the lack of any outward trappings of mathematics, were considerable. Readers and reviewers noted that the text was dense with abstruse terms such as 'apsides', 'ellipse', 'cosine', 'isogeothermal', 'perturbations', and 'eccentricities'. These hard words were defined in an extensive glossary, but their frequency and the density of the argument made the work a tough if worthwhile read. In one of the few negative reviews, the Anglican religious quarterly *British Critic* complained that 'to the beginner in science it would be perfectly unintelligible, and as a digest for one more advanced it is incomplete'.[30] There was no overall narrative structure, and the first edition lacked even a table of contents. The difficulties were most acute in the sections on experimental sciences such as light, electricity, magnetism, and sound, for these subjects did not have the kind of unifying explanation that gravity provided for planetary and stellar astronomy. Rather, the links work at the level of sentences, paragraphs, and sections, as *Connexion* weaves together specific threads between areas of enquiry.[31] Thus, knowledge of the principles of matter is essential for understanding its effect on light, so that we can then predict the path of light through the atmosphere. Using that knowledge, the velocity of light can be calculated through the study of the eclipses of Jupiter's satellites, which in turn makes it possible to use stellar parallax to determine the real motion of the earth in space. Each subject, traced in its ramifications, opens up prospects for understanding the others. These links create a sense of expectation and suspense, in which the processes, discoveries, and debates of science reveal unexpected points of contact between different bodies of knowledge. The overwhelmingly positive reception of *Connexion* resulted from the realization of readers and reviewers that it was not a primer, but an attempt to impart a more comprehensive understanding of the connecting links between different scientific findings than the study of any one discipline could impart.

The grandeur of the subjects being treated gives an immediate sense of the sublime, and many passages can be extracted from the work that are simultaneously accessible and impressive in their descriptive power. But most contemporary readers recognized that this was no facile treatise on the wonders of nature, as the explanation of sublimity is achieved only through a high level of abstraction. A contrast between the quiet beauty of the sunset and the dazzling, if not blinding, effect of the midday sun leads to an explanation of how the diminished light and heat of the sun are caused by the power of the atmosphere to absorb light. As elsewhere in *Connexion,* the rewards come through careful reading and full mastery of each step of the argument:

> The stratum of air in the horizon is so much thicker and more dense than the stratum in the vertical, that the sun's light is diminished 1300 times in passing through it, which enables us to look at him when setting without being dazzled. The loss of light, and consequently of heat, by the absorbing power of the atmosphere, increases with the obliquity of incidence. Of ten thousand rays falling on its surface, 8123 arrive at a given point of the earth if they fall perpendicularly; 7024 arrive if the angle of direction be fifty degrees; 2831 if it be seven degrees, and only five rays will arrive through a horizontal stratum.[32]

The reader is asked to inhabit a geometrical framework, dominated by a contrast between looking straight across the horizon and vertically up to the sky. The initial statement ('the sun's light is diminished 1300 times') contrasts the effect of the atmosphere at noon and the end of the day in simple numerical terms; but the reader must then interpose a series of intermediate steps between these extremes, as the angle of incidence of the light in relation to the horizon gradually becomes more oblique. Sunlight can be broken down into rays, and these can be counted. The reader is provided with a striking demonstration of the light-absorbing powers of the air, as 'ten thousand rays' are reduced to 'five'.

This discussion of what is observed then moves to the consequences for human observers and perceptions of nature. Much of what we see is an optical illusion:

> Since so great a quantity of light is lost in passing through the atmosphere, many celestial objects may be altogether invisible from the plain, which may be seen from elevated situations. Diminished splendour and the false estimate we make of distance from the number of intervening objects, lead us to suppose the sun and moon to be much larger when in the horizon than at any other altitude, though their apparent diameters are then somewhat less.

But the aim of understanding physical truth is not to diminish nature by showing that all is a meaningless optical trick due to humans being misled by their senses, but rather to heighten the beauty of what we can see:

> Instead of the sudden transitions of light and darkness, the reflective power of the air adorns nature with the rosy and golden hues of the Aurora, and twilight. Even when the sun is eighteen degrees below the horizon, a sufficient portion of light remains to show that, at the height of thirty miles, it is still dense enough to reflect light. The atmosphere scatters the sun's rays, and gives all the beautiful tints and cheerfulness of day. It transmits the blue light in greatest abundance; the higher we ascend, the sky assumes a deeper hue, but in the expanse of space, the sun and stars must appear like brilliant specks in profound blackness.[33]

Accounts of 'the rosy and golden hues of the Aurora' and 'the cheerfulness of day' call up images that would have been familiar from contemporary paintings and descriptive verse. They were also becoming staples of other emerging genres of science writing, as in Davy's *Consolations in Travel.* But in *Connexion,* such commonplaces appear in a different light—quite literally—because they alternate with references to the angular position of the sun and the diffusing power of its rays. Emotive and aestheticizing phrases are set in a context of ordered perspectives and rational understanding. The text is not easy reading, and at least four words in this paragraph ('altitude', 'incidence', 'stratum', and 'vertical') appear in the glossary, but the persistent reader is rewarded with a sublime perspective on the great scheme of creation.

Emblems of Omniscience

Understanding the world of light from an abstracted geometrical space, the reader of *Connexion* begins to understand why we see the world around us in the way we do. The mathematical framework created in the text allows the reader constantly to shift perspectives, to see the stars from a position in the heavens, or to imagine how long it would take the inhabitants of the newly discovered asteroid Pallas to travel around their world in a steam-carriage. In the book's description of the telescopic surveys of John Herschel and his father William on distant galaxies and double stars, the most modest rhetorical means are employed to take the reader on an extraordinary virtual tour of the cosmos:

> So numerous are the objects which meet our view in the heavens, that we cannot imagine a part of space where some light would not strike the eye;—innumerable stars, thousands of double and multiple systems, clusters in one blaze with their tens of thousands of stars, and the nebulae amazing us by the strangeness of their forms and the incomprehensibility of their nature, till at last, from the imperfection of our senses, even these thin and airy phantoms vanish in the distance. If such remote bodies shine by reflected light, we should be unconscious of their existence; each star must then be a sun, and may be presumed to have its system of planets,

> satellites, and comets, like our own; and, for aught we know, myriads of bodies may be wandering in space unseen by us, of whose nature we can form no idea, and still less of the part they play in the economy of the universe . . .[34]

By the simple device of inviting the reader to 'imagine a part of space', *Connexion* opened a prospect of the universe from the perspective of the stars, thereby rendering superfluous the fantastical dream-mechanics in *Consolations in Travel.* The novelist Maria Edgeworth had complimented her friend on the simplicity of style, which avoided the 'ornaments of eloquence dressing out a sublime subject'. In Edgeworth's view these were 'just so many proofs either of bad taste in the orator, or of distrust and contempt of the taste of those whom he is trying thus to captivate'.[35]

For those hoping simply to master facts and theories, as the *British Critic* had complained, *Connexion* was 'incomplete' and 'unintelligible'. The pre-eminently skilled popularizer Harriet Martineau, for one, found it 'cumbrous and confused'.[36] Even more favourably inclined critics found Somerville's writing required close concentration. The seventy pages of Somerville's 'Preliminary Dissertation' had left Edgeworth 'long in the state of the boa constrictor after a full meal', her mind distended by the concentrated knowledge that the book had released.[37] Readers of *Connexion* were encouraged to approach the text intensively, in the way that had been characteristic of the period before the late eighteenth-century boom in the number of publications had encouraged rapid scanning and skimming, rather than intensive study of a few. There were no illustrations in the first edition, but Somerville's writing demanded that the reader follow the words on the page as carefully as the points and lines of a geometrical diagram.

This method of close reading had been traditionally applied to the Bible and other religious texts, and it is no coincidence that the physical world of *Connexion* is pervaded by a sense of the divine:

> The heavens afford the most sublime subject of study which can be derived from science. The magnitude and splendour of the objects, the inconceivable rapidity with which they move, and the enormous distances between them, impress the mind with some notion of the energy that maintains them in their motions with a durability to which we can see no limit. Equally conspicuous is the goodness of the great First Cause, in having endowed man with faculties by which he can not only appreciate the magnificence of His works, but trace, with precision, the operation of his laws; use the globe he inhabits as a base wherewith to measure the magnitude and distance of the sun and planets, and make the diameter of the earth's orbit the first step of a scale by which he may ascend to the starry firmament. Such pursuits, while they ennoble the mind, at the same time inculcate humility, by showing that there is a barrier which no energy, mental or physical, can ever enable us to pass: that however profoundly we may penetrate the depths of space, there still remain innumerable systems, compared with which those apparently so vast must dwindle into insignificance, or even become invisible; and that not only man, but the globe he inhabits,—nay the whole system of which it forms so small a part,—might be annihilated, and its extinction be unperceived in the immensity of creation.[38]

The most obvious and immediate significance of *Connexion* was theological. Through framing passages stressing the devotional relevance of the physical sciences, Somerville showed that the higher mathematical analysis of the French need not lead to atheism or indifference (recall that, notoriously, the *Mécanique céleste* had never mentioned God), but could be used to formulate a more profound understanding of the manifestations of divine creation. Unlike the Bridgewater Treatises and other contemporary works on natural theology, however, *Connexion* pays almost no attention to particular adaptations in nature as evidence of God's goodness and power, and focuses instead on the general laws governing matter. The connected explanation of these laws provided by science is not seen as support for a thoroughgoing materialism, as it had been by Laplace, but rather as evidence of God's all-knowing foresight.

The outstanding accomplishment of contemporary science in understanding these laws, as presented in *Connexion,* was the discovery of the ether. The ether was described as a weightless, invisible substance occupying the interstices between the atoms of ordinary matter, from the farthest reaches of interstellar space to the living bodies of animals and plants. The great power of the ether was in providing the necessary physical substrate for understanding the material world, notably in the case of light, but potentially for electricity, magnetism, heat, and other forces. In much the same way that the atmosphere carried waves of sound to our ears, *Connexion* explained, the ether transmitted waves of light to our eyes.[39] The ether then carried the impression of these waves through our eyes to our nerves, ultimately to serve as the basis for comprehension. Somerville thought the transmission of light through the ether occurs every time we read: we can understand the text on the pages of *Connexion* because our bodies, like the rest of the physical universe, are pervaded by the ether.

> This study teaches us that no object is seen by us in its true place, owing to aberration; that the colours of substances are solely the effects of the action of matter upon light, and that light itself, as well as heat and sound, are not real beings, but mere modes of action communicated to our perceptions by the nerves. The human frame may therefore be regarded as an elastic system, the different parts of which are capable of receiving the tremors of elastic media, and of vibrating in unison with any number of superposed undulations, all of which have their perfect and independent effect. Here our knowledge ends; the mysterious influence of matter on mind will in all probability be for ever hid from man.[40]

An understanding of the role of the elastic ether, however important for Somerville, clearly had limits, for if everything could be explained by the laws of matter in motion, the result would be pure materialism. As this passage makes clear, the human mind was likely to remain outside the bounds of comprehension, although in allowing us to understand the laws of the physical world, mind was the greatest of all divine gifts.

Connexion closes with a declaration of faith that mathematics is the highest form of theology. The closest approach the reader could make to understanding God's laws was through mathematics, an essential element of the human mind implanted by God when he created Adam:

> These formulae, emblematic of Omniscience, condense into a few symbols the immutable laws of the universe. This mighty instrument of human power itself originates in the primitive constitution of the human mind, and rests upon a few fundamental axioms which have eternally existed in Him who implanted them in the breast of man when He created him after His own image.[41]

Behind these words was a profound personal commitment. Somerville had been brought up in the stern traditions of Scottish Presbyterianism, with its stress on original sin and damnation. She rebelled against this at an early age, and her most powerful experiences as a child involved a sense of direct contact with nature, which allowed her to escape into an inner world free of restraints. There is a strong sense that Somerville felt most intensely alive and completely herself in the pursuit of mathematics. The divine transcendence of God's power, she believed, could most fully be experienced by those who understood mathematical language.

It was this divine spark which Somerville believed she had felt as a young woman when she first learned about algebra, and that those around her had attempted to stamp out. Her confidence in a divinely implanted harmony of understanding led her to believe in the right of women to achieve their full potential; even as a child she thought it 'unjust that women should have been given a desire for knowledge if it were wrong to acquire it'.[42] She never could find God in formal church-going, but instead traced his hand in those 'few fundamental axioms' that she had discovered through her self-education in mathematics. On religious matters, her father-in-law and uncle, the Rev. Dr Thomas Somerville of Jedburgh in Scotland, became her confidant. As she explained in writing to him from London, she never felt comfortable with the forms of the state churches either in Scotland or England, and used the example of her daughters Margaret, Mary, and Martha to illustrate the logic of Unitarian views.

> I attended the established church; but with little satisfaction; shocked by their creeds and struck with the Litany addressed almost *exclusively* to our saviour.... I dont believe in the Trinity at all, but I imagine our saviour to be sent by the one God to instruct us in his will, now I cannot tell whether he was really a mere man, or if he existed before he came into the world.... The next point is whether Christ died as an expatiation for our sins, or only as a martyr to prove the truth of what he taught[.] I dont think original sin a doctrine at all consistent with the goodness of God, and if there is no original sin what need of expatiation?... The last point and not the least important is how to instruct the children[.] Margaret said the other day when we were not thinking on religion, Papa how is it that God and Christ and the Holy Ghost are not three but one you may as well say that Martha and Mary and I are not three daughters but one. I had never mentioned the subject to her of course.[43]

To discuss such opinions, as Somerville well knew, 'would make people say I had become an infidel'. Her sympathies lay with the rational religion of the Unitarians, who stressed the power of God but did not believe in Christ's divinity. She and William occasionally attended a local Unitarian chapel. But this had to be done in secret, as friends had 'repeatedly declared before me that nothing on earth w[oul]d enduce them to sit in the same room with a Unitarian or hold any correspondence with them, and that it is highly criminal to go even to hear them'.[44] In the end, she followed her uncle's advice: the family discussed their real views at home, but rented a pew in the local Anglican Church. Church-going was important in avoiding both the appearance and reality of outright scepticism. Somerville may have abandoned traditional Christianity, but she had replaced this by a passionate faith in a God who could be best understood through mathematics.

Behind the appearance of conformity, *Connexion* thus embodied religious views that were widely despised. Reviewers voiced suspicions.

The *British Critic* was struck by the stress on material laws and the lack of any reference to Christ's death on the cross—the doctrine of the atonement for human sin that was at the heart of true Christianity. *Connexion,* the reviewer admitted, was not alone in these failings, which were also to be found in Herschel's *Preliminary Discourse* and other contemporary works of physical science. Why were religious references in such treatises confined to a few opening and closing remarks, rather than being reinforced through 'reiterated and glowing allusions' to Christian doctrine? Why did the 'connexions' discussed by Somerville not encompass theology, which after all had traditionally been viewed as 'queen of the sciences'?[45] Most reviewers, however, considered the references to a creator God as sufficient.

Whatever doubts there might be about the orthodoxy of *On the Connexion of the Physical Sciences,* its author would not suffer the fate of Hypatia. As recounted in the *Quarterly Review* (quoting from Edward Gibbon's *Decline and Fall of the Roman Empire*), Somerville's celebrated classical predecessor in mathematical learning was 'torn from her chariot, stripped naked, dragged to the church, and inhumanely butchered; her flesh was scraped from her bones with oyster-shells, and her quivering limbs were delivered to the flames'.[46] Merely alluding to Hypatia, however, indicated that a woman mathematician was a powerful and dangerous symbol. For conservatives like the religious reviewer in the *British Critic,* Somerville needed to embody not only learning but the virtues of spotless morals and evangelical piety; for Brougham and the advocates of self-education, her status as a clever, competent woman was enough.

Somerville's situation was widely cited in the later phases of the debates over the decline of science. In a review of *Connexion,* David Brewster regretted what he interpreted from his very partial perspective as yet another example of an author turning away from original discovery to write for popular approval, while others saw Somerville's very existence as proof that the declinists were wrong.[47]

A marble bust of her was commissioned for the rooms of the Royal Society, paid for with subscriptions from reformers and anti-reformers alike, and she was made an honorary fellow of the Astronomical Society. As part of an effort to embarrass the Whigs and recognize scientific accomplishments in the wake of the declinist controversy, Sir Robert Peel's Tory administration awarded her an annual civil list pension of 200 pounds in 1835, the year following publication of *Connexion*. Within the family, the failure of the preceding Whig government to provide such rewards, despite the Somervilles' long record of support for the party, was despised; Brougham had served as Lord Chancellor for four years and was supposedly a friend, but had never lifted a finger. As Somerville's son wrote to his mother, 'will that ungentlemanlike scoundrel Brougham *ever dare* to look you in the face again?'[48] Somerville knew her value as a symbol.

All reviewers of *Connexion* were acutely aware of just how unusual it was for a woman to provide a mathematically grounded synthesis of the physical sciences. They tended, however, not to use Somerville to add the physical sciences to the roster of 'female accomplishments', but to stress that she was unique, an exception to jibes about learned ladies and bluestockings. Herschel, writing anonymously in the *Quarterly Review,* could find none of the stereotypical signs of a feminine hand in the *Mechanism of the Heavens,* and remarked upon the 'entire absence of anything like female vanity or affectation', claiming that the author 'seems entirely to have lost sight of herself'. 'Mrs Somerville,' the *Edinburgh Review* similarly noted of *Connexion,* 'is the only individual of her sex in the world who could have written it'. The *Printing Machine* noted that a prejudice against learned women seemed to be a western preserve, unlike attitudes to be found in India and the Arab world.[49]

If it was thought that few women in history possessed the requisite mathematical understanding to write such works, those few could also be considered capable of developing the unifying perspectives they demanded. *Fraser's Magazine* flippantly attributed this to the 'quick-sighted' eye for the main chance by 'female aspirants for literary reknown', comparing 'the transcendent blue light of Mrs. Somerville, F.R.A.S.' to the failed attempts of Dionysius Lardner, editor of the *Cabinet Cyclopaedia*, at authorship. 'How much superior is the learned lady to the discomfited quack, and how well does she wear her laurels!'[50]

This was satire, but the Cambridge polymath William Whewell also indicated that Somerville had the kind of mind best suited to tackling the deeper problems of philosophy. In his view, the precision of the language and structure of *Connexion* had everything to do with its authorship by a woman. If men were active, prone to confusing practice and theory, women were above the fray, their reasoning unclouded by practical consequences, giving their perceptions clarity and their writing transparency. On Whewell's reading, female authorship offered the possibility of direct and detached insight into the laws of nature. The result was an elevated perspective achieved not through hard labour or conquest (as in the works of the great travelling naturalists such as Alexander von Humboldt), but through inner illumination, because nature itself was gendered female.[51] The review quoted a celebrated passage from the poet John Milton, who had described such a viewpoint in his early play *Comus:*

> In regions middle, of calm and serene air,
> Above the smoke and stir of this dim spot
> Which men call earth.[52]

Not everyone shared Whewell's Olympian perspective on Somerville's accomplishments as a woman. Radical parliamentarians, raising her case in the House of Commons, saw nothing transcendent about the 200 pounds civil list pension awarded by Peel, which they condemned as a waste of precious funds at a time when the poor were starving. The Whig member of Parliament for Cornwall, Charles

Buller, pointed out that neither *Connexion* nor *Mechanism of the Heavens* were works of original discovery, and that Somerville was a wealthy woman 'in easy circumstances'. Why should authors of such books, however lofty their subject matter, be awarded public funds to enable them to live off 'the highly-taxed industry of the labouring poor'? Some of the army pensioners at Chelsea, where William Somerville served as physician, were overheard wondering why his wife should be awarded 200 pounds a year for being clever and 'for writing a book that no one can understand'.[53]

The Empire of Analysis

Walking on a Cornish beach in the early 1850s, a passer-by found a wooden plank emblazoned with the words 'Mary Somerville'—a fragment of a ship lost at sea on its homeward voyage from Calcutta. Nearly two decades earlier, a wealthy Liverpool shipbuilder had asked the author of *Connexion* if he could name a merchant vessel in her honour and use a replica of Francis Chantrey's recently completed bust for the rooms of the Royal Society as a figurehead. Somerville gave the shipbuilder a copy of her newly published second edition, and in return he gave her a down tippet, or scarf, recently arrived from Bengal.[54] Until its shipwreck, the ship carried cotton, tea, and flour on the trade routes from Liverpool to India and China.

As the daughter of a naval hero in the wars with the French, Somerville had been delighted to become, in a very literal way, a symbol of overseas trade. At a practical level, her writings were closely tied to the imperial ambitions of the British economy. *Connexion* dealt with issues of global direction-finding central to the great 'Magnetic Crusade' that was just getting underway through the auspices of the British Association for the Advancement of Science, and the *Mechanism of the Heavens* had introduced mathematical solutions vital for geodesy, surveying, and map-making. Both books would contribute to placing the physical sciences at the leading edge of the push towards empire. Above all, they encouraged readers to take a global perspective, to see the earth as an object of geometry and logic, comprehensively and as a whole. As *Connexion* noted in its closing pages, 'The common bond of analysis is daily extending its empire, and will ultimately embrace almost every subject in nature in its formulae.'[55] This imagined future revived the notion of *mathesis universalis*—René Descartes' seventeenth-century dream of a complete understanding through mathematics. But that mathematizing vision was now embedded in a belief in commercial progress and imperial expansion, made possible by science and underwritten by a profound faith in the virtues of contemporary civilization. Civilization was steam, travel, education, printing, and the rapid advance of the arts and sciences. It was the triumph of learning over barbarism, of science over superstition, of the 'refined' over the 'savage' races. It was everything Somerville's own accomplishments and story of self-education could teach readers. It was a world in which women, through the power of knowledge, might take their rightful place.

5

A Philosophy for a New Science

Charles Lyell's *Principles of Geology*

> The great merit of the *Principles* was that it altered the whole tone of one's mind, & therefore that, when seeing a thing never seen by Lyell, one yet saw it partially through his eyes.
>
> —Charles Darwin

Of all the new sciences, among the most contentious was geology. Forged from eighteenth-century mineralogy, antiquarianism, theology, and mining lore, it was often seen as speculative or even as undermining religion. It was with the aim of establishing secure foundations for the science that Charles Lyell, an ambitious young lawyer in London, wrote the *Principles of Geology*, published in three volumes between 1830 and 1833. This was a book about seeing: using what people could observe around them to penetrate the unseen and invisible world of the deep past. *Principles* replaced the spectacle of geological catastrophe, in which earth history was punctuated with episodes of violent change, with a vision of slow processes, acting over untold eons of time. Many readers used the book to come to terms with the consequences of scientific findings in relation to the biblical accounts of the Creation and the Flood; it also shaped Darwin's scientific vision as he set off in 1830 to circumnavigate the globe on the *Beagle.*

The title proclaimed heroic ambitions—no less than doing for the study of the earth what Isaac Newton had done for astronomy and natural philosophy in the *Principia Mathematica.* But unlike Somerville's *Connexion, Principles* aimed to do this not through a literary version of mathematical reasoning, but by understanding causes. Those studying the history of the earth, Lyell's book argued, should carry out their investigations under the assumption that causes now visible (volcanoes, rivers, tidal currents, earthquakes, storms) are of the same kind that have acted in the past, and have done so with the same degree of intensity as in the present. In the *Principles,* this assumption of uniformity did not dictate a particular narrative history of nature, but instead offered a policy for securing the philosophical foundations of geology: the book aimed to define, as Lyell told a friend, the 'principles of reasoning in the science'.[1]

Principles attracted a wide range of readers in Europe, America, and Australia, from explorers and engineers to poets and artists. Among its most influential advocates was John Herschel, who made much of the book's stress on visible causes in his *Preliminary Discourse,* which was published only a few months after Lyell's first volume. *Principles,* celebrated for its startling analogies and the elegance of its prose, highlighted issues of universal decay and the vastness of time which entered into Alfred Lord Tennyson and Matthew Arnold's poetry, and contemporary novelists used the book to come to terms with contingency, change, and transformation. The atheist agitator Charles Southwell, in prison in 1842 for blasphemy, asked for a copy along with his accordion and some cigars.[2]

Reforming the Earth

In 1830 geology was lively, popular, and controversial, a product of the transformation of the study of nature in the early decades of the nineteenth century. The Geological Society of London, founded in

1807, had emerged as one of the most successful of the new specialist associations that had been so bitterly opposed by Joseph Banks. In these early years, geology engaged a network of practitioners ranging from physicians and aristocrats to engineers and farmers. A focus on strata gave these diverse constituencies a practical programme of research and a common goal. Maps showing the distribution of the strata were vital to mining, agriculture, and a philosophical understanding of earth history.[3]

In being open to such a wide range of participants, however, geology was in danger of gaining a reputation for philosophical promiscuity. Underlying disagreements about theological issues could erupt into speculative excess or religious scepticism. Despite the focus on strata, public debate about the meaning of geology remained embedded in controversy about Creation, the Fall, and the Flood. In this context, any argument for the constancy of natural laws—as opposed to a providential nature dependant upon divine intervention—was bound to create controversy. Who was to interpret the meaning of a science whose findings could so flagrantly be used to contradict the opening verses of the Bible? How was a scientific view to be given of the history of life which did not lead to soul-denying materialism and atheism? Writing in these circumstances involved tactics, presenting science in a way that would calm fears of irreligion. This was a common problem, similar to those faced by Mary Somerville in transforming Laplace's godless celestial dynamics into the *Mechanism of the Heavens*. It required creating a new image of geology and conveying that image effectively to readers. The story of the *Principles* is thus one of strategies and silences, of the differences between writing for a public and views expressed only in private letters and conversations with friends.[4]

The author of the *Principles* was born in 1797, and like his contemporaries grew up under the shadow of the French Revolution and fears of foreign invasion. Lyell's father was a wealthy gentleman and opponent of reform, who had inherited a large estate in Forfarshire in Scotland but preferred to live in a rented estate in southern England. The young Lyell grew up in a cultured environment typical of the moderate Tory gentry, who aimed to adapt the best traditions of Church and State to changing circumstances. When Lyell entered Oxford University at the age of seventeen, classics and theology dominated the curriculum, as might be expected at an institution devoted to the Establishment. Lyell read classics and his writing received sufficient praise for him to think of winning prizes and becoming a poet. He also attended lectures on geology, having already developed an interest in natural history through browsing in the library of his father, who was a keen botanist. Geology had been introduced at Oxford to arm undergraduates against the infidels, and like other natural sciences it was an extracurricular option which did not lead to a degree. Lyell attended the flamboyant lectures of the Rev. William Buckland, whose daring reconstructions of extinct monsters and lost worlds attracted an enthusiastic following. Buckland stressed ties to classical learning and agricultural utility: and because many feared that the new science undermined the truth of the Mosaic narrative in Genesis, he contended that geology evidenced divine design and a universal Deluge. The result was a romantic vision of the progress of life through countless ages, populated by strange animals perfectly adapted to even stranger physical conditions, and culminating in the creation of the human race.

By the early 1820s, Lyell had graduated and moved to London in preparation for a career as a barrister. However, poor eyesight and ambitions to shine in literary circles led him to shift his attention to science, despite his father's worry that he was abandoning a secure profession. In making geology his vocation—while also hoping to make money from it—Lyell was doing something new. In the early nineteenth century, most activities that might be called professional in science (other than university professorships) were seen as low status, involving specimen-selling, instrument-making, curating collections, and hack writing. Lyell, like others at this time, hoped to raise authorship into a calling fit for gentlemen. At the same time, as John Herschel would show through the *Preliminary Discourse*, gentility itself was to be redefined around notions of intellectual leadership.

Living in the metropolis Lyell associated with 'Lawyers, Geologists & other sinners'[5] and moved away from the moderate Toryism of his family. By the mid-1820s he had become an ardent liberal Whig, advocating electoral reform and disestablishment of the Anglican Church. His experience of the ancient universities, which he thought were in a bad way, was instrumental in his change. To his mind only reform could salvage the established order from the horrors of 'mob-rule'. Although his privileged background shielded him from the rawer aspects of day-to-day political struggles, they could not be ignored. Lyell witnessed violent revolutionary demonstrations in Paris, and at home in Forfar there were threats that if the House of Lords rejected reform, the mob would smash the Tory carriages for a country ball.[6] The most immediate (although less violent) conflicts that Lyell faced were within his family, for his father and brothers had no sympathy for reform. At a crucial by-election in Forfarshire in 1831, when the rest of the family canvassed for the Tories, Lyell confronted a difficult choice. He could not violate his beliefs by voting Tory, but neither could he slight his father by siding with the Whigs. At the height of the Reform Bill agitation and in an open election with fewer than a hundred voters, Lyell abstained.[7]

Anxious to avoid such conflicts in the future, Lyell devoted himself to science, which he saw as a refuge from political and religious strife. 'As for public affairs,' he told his fiancée, Mary Horner (herself from a famous family of reforming Whigs), 'I have long left off troubling myself about them, as knowing that one engaged in scientific pursuits has as little to do with them, in point of influencing their career, as with the government of hurricanes or earthly motion.'[8] For Lyell, science offered an indirect way of forwarding reform without betraying his father or his teachers through outright rebellion. Since a gentleman's right to independent judgment was acknowledged, at least in principle, he remained welcome both at home and in fashionable Tory circles in the metropolis. The centre of his social life was the Athenaeum, a club for literary, scientific, and artistic gentlemen formed in 1824. It grew out of meetings in the Albemarle Street premises of the publisher John Murray, and fostered the dominance of British intellectual life by clubbable gentlemen.

Lyell put his family's Tory connections to strategic use, taking the battle for reform into what he now privately saw as the enemy camp. When he briefly assumed an academic chair as part of his campaign to become a new kind of professional man of science, it was at King's College, founded by Anglican Tories to counter the utilitarian London University. Lyell reviewed for the Tory *Quarterly* not the Whig *Edinburgh*, where reforming views on education and science had been trumpeted for decades. He brought out his book with Murray, a Tory, not with the useful knowledge merchant Charles Knight. Lyell of course knew about a much wider spectrum of political and religious groups, from radical socialists to the millenarian followers of the Rev. Edward Irving. But the first edition of the *Principles* was not written for any of these; instead, it targeted a conservative and respectable readership, made up of gentlemen and ladies who feared that geology was anti-Bible and anti-Christian, and needed to be convinced that science had nothing to do with materialism. In the first instance he hoped to reach the gentry, aristocracy, and professional classes who might be familiar with his early reviews. Behind the scenes Lyell manoeuvred to get sympathetic reviewers for his book in the *Quarterly Review* and the religiously orthodox *British Critic*. 'It is just the time to strike,' he told his ally George Poulett Scrope, 'so rejoice that, sinner as you are, the Q. R. is open to you.'[9]

One of the greatest assets of the *Principles* was its publisher. Murray produced the first two editions on fine paper, the three octavo volumes together costing two pounds and five shillings, making it affordable only by the wealthy and influential—precisely the market Lyell hoped in the first instance to address. He regretted not following expert advice about the type, which 'old people' found too small and which misled potential polite readers into thinking it was more of 'a hard-meat, business-like book of mere science than it really is when entered on'.[10] Despite these obstacles, the book sold well, so that the 1,500 copies of the first edition of volume one were soon exhausted.[11]

The early editions of *Principles*, in the form published by Murray, thus aimed to unite an enlightened clerisy of truth-seekers who would diffuse knowledge to the 'vulgar'. This elitist, authoritarian vision of 'popular science', increasingly dominant from the 1830s, contrasted with older traditions that had encouraged participation by a wide range of people in the making of knowledge. Young Lyell, like Herschel, Babbage, and Whewell, tended to dismiss provincials, colonials, women, and working people as sources of information in an intellectual hierarchy, whose facts gained meaning by being drawn together and reasoned upon by the few.[12] Irate colleagues in America, tired of an itinerant bigwig plying them for hard-won findings, later nicknamed him 'The Pump'. Lyell saw no need for ordinary readers to master all the research and reasoning that had gone into the making of knowledge; rather, experts should 'communicate the results, and this we are bound to do'.[13] Only if a science could be presented as gentlemanly, disinterested, and apolitical could it take its place as a force for reform.

Observing Causes

Published at a pivotal moment in the public history of the sciences in Britain, the *Principles* was the most thoroughgoing vindication of geology as a sophisticated and philosophical enquiry. More than any of his associates, Lyell worried that an impressive edifice of practical research into the earth's strata could be accused of resting on shaky foundations. In the specific sense of research conducted according to clearly articulated rules of reasoning (such as those in Newton's *Principia*), he believed that geology was not a science. His ambition was, single-handedly, to make it one.

What did it mean to attempt to make a subject into a science? Most geological works offered a connected narrative of progress similar to that narrated by the Unknown in *Consolations in Travel*, or they were historical in the older, descriptive sense of 'natural history', outlining the order of the strata. In contrast, the *Principles* was a treatise on natural philosophy: what the subtitle of the work referred to as 'an attempt to explain the former changes of the earth's surface, by reference to causes now in operation'. The key word here was 'causes'. The *Principles*, as Herschel's *Preliminary Discourse* would point out soon after the first volume was published, depended on the notion of a *vera causa* or 'true cause', defined as a cause in operation which could be *observed*. As we have seen, the notion of a *vera causa* had been developed in Newton's *Principia* and in Scottish common sense philosophy.[14]

Three rules of reasoning underpinned the proposals in the *Principles* for reforming the perceptions of readers. First, geologists must assume that the basic laws of nature (such as gravity) had not changed over time. The sole exception was the origin of new species, where the text was non-committal, claiming that the human span of observation had been too short. Even here, though, it left open the possibility of an explanation through natural law, indicating that there might be a middle way between creation and evolution. Since, however, there was no observable basis for theorizing on this issue—no *vera causa*—the *Principles'* genteel silence on these alternatives appeared consistent with its general philosophical position. Almost all geologists accepted the constancy of the laws of nature, an assumption that directly tied the *Principles* to Newton's *Principia*.

The second proposed rule of reasoning involved the kinds of causes to be used in interpreting the earth's past. Here the *Principles* took a far more controversial stance, arguing that no causes other than those that readers could see acting should be employed in explanations. Of course, different causes from those occurring now might be responsible for the past condition of the earth; almost all practitioners thought that this was the case. But the *Principles* argued that such causes could be part of science only if they were observed by reliable witnesses. Recourse to such causes, without the backup of trustworthy testimony, encouraged investigators 'to cut, rather than patiently to untie, the Gordian knot'.[15] Critics complained that this was unduly restrictive: why assume that human observation was the only measure of untold eons of earth history?

Most surprising of all, the *Principles* argued that interpretation must proceed on the basis that causes had not varied in degree either; in short, the intensity of observable earthquakes, volcanoes, floods, and other geological agents should be the measure of their action in the past. Again, knowledgeable commentators such as Adam Sedgwick, professor of geology at Cambridge, pointed to twisted mountain strata and giant erratic boulders as good evidence for episodes of 'feverish spasmodic energy' against which modern causes paled into insignificance.[16] Quiet stasis, in this view, was punctuated by violent catastrophe and sudden death. But according to the *Principles,* recourse to such unobserved events had made geology a byword for unrestrained speculation. Readers could see for themselves, it claimed, forces adequate to produce the Andes and the Alps, especially if the cumulative effects of time were taken into account.

The book developed the implications of this method in opposition to some of the best-supported generalizations of contemporary geology. Lyell's early periodical reviews had supported a progressive history of life, beginning with invertebrates and followed by fish, reptiles, mammals, and then man. But the *Principles,* in contrast, argued that this was an unphilosophical assumption which ignored the patchy nature of the fossil evidence. Where most geologists dreamed of filling in the lost page in the book of life, the *Principles* suggested that they should work under the assumption that almost the entire volume had been destroyed. Gaps in the record, such as that between the Cretaceous and Tertiary, might represent vast ages, not a sudden catastrophe. Similarly, the absence of the remains of mammals, birds, and other complex forms from the very oldest rocks did not necessarily mean that such creatures had not lived then, but only that their remains might not have been preserved. This suggestion was not unreasonable, as the discovery of a fossil opossum—described in 1824—from the so-called 'Age of the Reptiles' had recently demonstrated that mammals had existed long before geologists had thought possible.

While denying any need to assume progress within the history of life, the *Principles* accepted that hotter temperatures had probably prevailed in the past. How was this to be explained while adhering to the philosophical rules of reasoning? Climate change, at least potentially, needed to be displayed as the result of the action of observable causes—not by a comet or a change in the earth's axis (a cause differing in kind) nor by a cooling earth (a cause differing in degree). The solution, given pride of place in the first volume, was one of the triumphs of the *Principles*. Deploying a geographical argument, the book showed how shifting relations between land and sea could produce the required changes (Fig. 14). Open ocean at the poles would result in a warmer climate, while elevated land masses there would produce cold. As Lyell told the surgeon Gideon Mantell, he had 'a receipt for growing tree ferns at the pole, or if it suits me, pines at the equator; walruses under the line, and crocodiles in the arctic circle'.[17]

With such arguments the *Principles* rejected the possibility of constructing any narrative 'story of the earth' at all; in this specific sense the book was profoundly ahistorical. The sole defining narrative event in the entire work was the advent of the human race. Within the

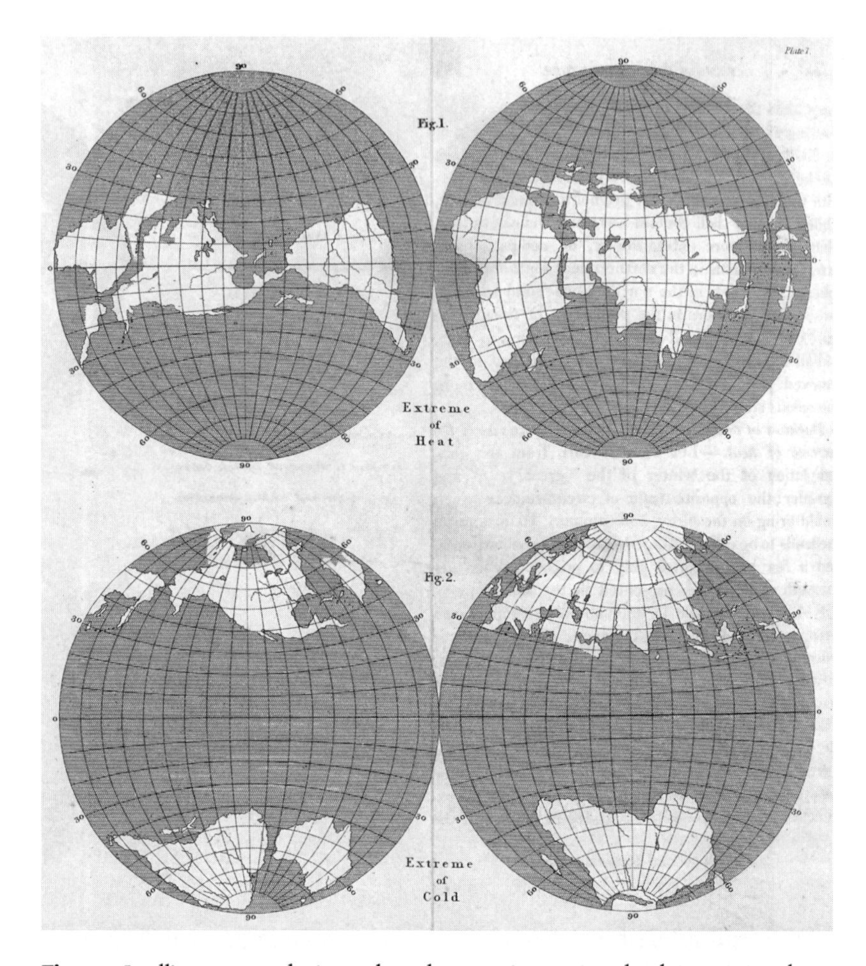

Fig. 14. Lyell's most audacious thought-experiment involved imagining how different configurations of land and sea could affect the climate. In this way the *Principles* combated assumptions that the earth as a whole had been cooling throughout its history, which Lyell saw as an unwarranted speculation. This engraved plate, added to the first inexpensive edition of *Principles*, made the point clear for a wide audience. 'Maps showing the position of land and sea which might produce the extremes of heat and cold in the Climates of the globe', Charles Lyell, *Principles of Geology*, 4 vols (London: John Murray, 1835), 1, plate 1.

random, shifting balance of forces, the uplift of mountains, and the erosion of coasts, the creation of humanity marked the beginning of one era and the end of another; for human origins were kept outside the physical flux that constituted earth history. Denying any shape to the records of the earth's deep past kept humans separate from the rest of creation and outside the reign of natural law, thereby underlining their special moral status. Although geology and history used analogous methods, the boundary between the two fields was sharply drawn: 'Geology is intimately related to almost all the physical sciences,' the *Principles* noted, 'as is history to the moral.'[18] Notably, human history provides the only sustained forward narratives in the *Principles,* and the 'historical sketch of the progress of geology' in the first volume and the history of Tertiary geology in the third are militantly Whiggish, developmental, and progressive.

The earth, on the other hand, exhibited no such signs of progress. The *Principles* underlined that change was not directional by speculating on the future, when a warmer climate might once again prevail. 'Then might those genera of animals return,' it predicted, 'of which the memorials are preserved in the ancient rocks of our continents. The huge iguanodon might reappear in the woods, and the ichthyosaur in the sea, while the pterodactyle might flit again through umbrageous groves of tree-ferns.'[19] Readers often understood this as involving a cyclical theory of earth history. A cartoon by Lyell's geological rival Henry De la Beche showed a future 'Professor Ichthyosaurus' lecturing to an audience of reptiles on the comparative anatomy of the extinct human race (Fig. 15).[20] However, the conditional 'might' in this famous passage ('Then might those genera of animals return') shows that there was no reason why such changes *must* or *will* come about. For all Lyell's romantic evocation of the return of the pterodactyls in a cycle of a 'great year', the substantive claims in the *Principles* were models of philosophical caution.

Fig. 15. Henry De la Beche's satire attributes a cyclical vision to the *Principles*, showing the return of ichthyosaurs and other reptiles from the distant geological past. A 'Professor Ichythysaurus' lectures on the fossilized remains of humans, who have become extinct. 'Awful Changes. Man found only in a Fossil State—Reappearance of Ichthyosaura', privately circulated lithograph (London, 1830).

This point needs to be stressed, for these statements about the pattern of earth history are often taken out of context, interpreted in terms of private letters and journals so that their function as thought-experiments about the past is obscured. Recommendations in the *Principles* about method have been turned into doctrines or 'isms' and compared with those of Lyell's contemporaries: Whewell began this process by dubbing Lyell a 'uniformitarian' in opposition to those, like Sedgwick, Buckland, and himself, who believed in 'catastrophism'. Such views make the *Principles* a cosmological book, which points towards the construction of a connected narrative history of the world. However, in his public statements during the 1830s Lyell no more advocated a steady-state, cyclical, or non-progressionist cosmology than he did progression itself. Indeed, the *Principles* claimed that *any* kind of global narrative would prove impossible to reconstruct, because too much of the record had been lost. Lyell was not the 'historian of time's cycle'.[21]

The application of basic philosophical principles in the book is thus almost entirely regulative: that is, geologists should carry out their investigations *as though* visible causes are the same kinds as those that have acted in the past, and of the same degree of intensity. Uniformity of law, kind, and degree had to be assumed, the book argued, to make geology scientific. The subtitle cautiously spoke of 'an *attempt*' to explain former changes; and the third and all subsequent editions softened this still further. In essence, the *Principles* aimed to make the study of the pre-human past amenable to the same rules of inductive inference that continental scholars, notably the great historian of Rome, Barthold Georg Niebuhr, were using to put the study of human history on a new basis.[22] The goal was to make geology into an inductive science grounded in the observation of causes. Lyellian uniformity was thus a method, not a doctrine about the shape of earth history.

All that readers needed to do, the *Principles* argued, was to reason on what they could see everywhere around them. The sheer length of the work—over 1,400 pages in the original edition—was essential to this programme of perceptual reform. Text and pictures, through the cumulative effect of hundreds of examples, made readers into witnesses to the power of modern changes. Over half the first edition is devoted to assembling a comprehensive toolkit of observed forces for use in interpreting the past. The descriptions of rivers, tides, currents, and delta formation were so extensive that the civil engineer Sir John Rennie used the *Principles* as a professional reference work while constructing waterworks on the Thames.[23] Aqueous causes were followed by a thorough account of their igneous counterparts. An entire chapter patiently documented the devastation wrought by earthquakes in one year in a single region in Italy. The second volume turned to the organic realm, discussing causes now in operation involving plants and animals. The main focus was on the relation between organisms and their environment, especially in relation to geographical distribution and impact of animals in altering the face of the globe. Long chapters were also devoted to the fossilization of organisms ranging from shipwrecked sailors to corals. This kind of detailed exposition occupied a central place in the book's strategy, for, as the philosopher Dugald Stewart had said, most people learned best from examples rather than general principles.[24] Even those who disagreed with the *Principles* admired its empirical richness, which mustered research from hundreds of sources in five modern European languages and two ancient ones. 'It is essentially,' the *Scotsman* newspaper noted, 'a book of facts.'[25]

The book drew extensively on French, German, and Italian sources, for, like Babbage, Herschel, and Somerville, Lyell knew that continental scientific research was often far ahead of the British. Much of the history was taken (with scant acknowledgement) from Giovanni Battista Brocchi's work on fossil conchology; information about inorganic causes derived from Karl Ernst Adolf von Hoff's massive compilation; Augustin de Candolle was the main source for plant geography; and Gérard Paul Deshayes provided tables of shells and assisted in establishing the innovative percentage system for dating strata used in the *Principles*. Lyell's greatest and earliest debt was to a group of geologists he had met in Paris, notably Constant Prévost. These men had shown, arguing against the celebrated naturalist Georges Cuvier, that the strata of the Paris Basin could be explained without reference to catastrophe, through comparisons with modern lakes, rivers, and seas.[26] The fieldwork described in the *Principles* also impressed readers, who gained a sense of direct contact with nature and were invited to picture an author climbing volcanoes, hammering chalk cliffs, measuring ancient temples, and excavating fossils with his own hands. A preface to the third volume stressed that travel had been central to the making of the *Principles*, with tours to France, Italy, and, above all, Sicily. Throughout the book, the vogue for 'Baconian' induction and empiricist theories of perception is clear.

The most obvious philosophical debt of the *Principles* was to Scottish traditions of theorizing about the earth. John Playfair, the Edinburgh professor who had advised Mary Somerville on her mathematical studies, had written *Illustrations on the Huttonian Theory of the Earth* in 1802. This work had brought the natural philosopher James Hutton's system of the earth, first published in 1788, into the mainstream of nineteenth-century debate, recasting it in inductive, empiricist terms acceptable to the post-Enlightenment era. Hutton had argued for a cyclical system of strata consolidation, mountain uplift, and decay, all produced by gradual forces operating under a beneficent Deity. Written from the perspective of a mathematician and natural philosopher, Playfair's *Illustrations* presented geology as a science which dealt with stable systems operating under unvarying laws. Lyell never seems to have read Hutton's original publications—his quotations were all at second-hand—but he used Playfair's works extensively.[27]

The philosophy of the *Principles* is often summed up, as is that of Hutton and Playfair, as 'the present is the key to the past'. But this slogan is unspecific and indeed almost meaningless: causal keys can function in many ways. For example, they can license reasoning by simple analogy. Rather the key worked because present causes could be—or had been—witnessed in action by reliable observers. It was the visibility of modern causes that made them the *only* legitimate basis of explanation. The *Principles* asked readers to imagine themselves as other kinds of beings, which would be better suited to underground exploration needed in geology: 'an amphibious being' would be better able to 'arrive at sound theoretical opinions' because so much of the earth is covered by water, while an underground gnome would be the best geological philosopher of all.[28]

But how could an observer, even a gnome, make inferences about the past? What, for example, could we legitimately infer about past climates from fossil remains? Recent geological discoveries had indicated that northern Europe had once been populated by hyenas, rhinos, elephants, and other creatures which today live in the tropics. Did this mean that the climate in these northern regions was formerly much hotter? Here again, as always, the *Principles* recommended relying only on what could be seen. The argument drew upon Dugald Stewart's *Elements of the Philosophy of the Human Mind,* published between 1792 and 1827. Stewart suggested that the observed continuity of animal instincts implied continuity of physical laws. According to the *Principles,* the existence of creatures found in a fossil state from a past period of earth history thus implied climatic conditions similar to those experienced by their counterparts in the present. 'Any naturalist will be convinced,' the *Principles* stressed, by the observation that 'the same species have always retained the same instincts, and therefore that all the strata wherein any of *their* remains occur, must have been formed when the phenomena of inanimate matter were the same as they are in the actual condition of the earth.'[29] It was this view of instinct, drawn from the most influential British philosopher of the early nineteenth century, which underpinned the *Principles*' central induction about the constancy of natural laws in relation to climate.

The emphasis on *vera causa* reasoning, the philosophical heart of the *Principles,* appealed to views about scientific method that were widely canvassed in the years of debate about reform. Lyell himself had little formal philosophical training and his references to philosophers are tactical and implicit. At Oxford he had been bored by formal logic and found Aristotle (whose writings still dominated the curriculum) 'an astonishingly stiff author'. But he would have known that Stewart and Reid were being discussed by authors across the religious and political spectrum. At Oxford, the moderate Tory Edward Copleston and his followers applied common sense philosophy as part of their project to create a Christian version of political economy.[30] After moving to London, Lyell frequently met John Herschel, whose *Preliminary Discourse* praised the *vera causa* theory of climate. In a letter to the author, Herschel welcomed the book as 'one of those productions which work a complete revolution in their subject by altering entirely the point of view in which it must henceforward be contemplated'.[31]

Few readers were that enthusiastic about *vera causa* explanations, but almost all agreed that the *Principles* had transformed the philosophical status of debate in geology. Writing anonymously in the *Quarterly,* Whewell celebrated the creation of a new science of 'geological dynamics': just as Galileo's work on motion had led to Newton's explanation of the planetary motion, so would the account of terrestrial forces in the *Principles* provide the basis for reconstructing earth history. Moreover, the work had proceeded 'not in an occasional, imperfect, and unconnected manner, but by systematic, complete, and conclusive methods'. Only in this way could geology 'be a Science, and not a promiscuous assemblage of desultory essays'.[32] This was high praise—even if Lyell was cast as the Galileo of geology and not its Newton.

Whewell's approval of the *Principles* was widely shared. At the second meeting of the British Association for the Advancement of Science in 1832, the geologist William Daniel Conybeare hailed the book (albeit before launching a critical barrage) as 'in itself sufficiently important to mark almost a new era in the progress of our science'. The *Monthly Magazine* called it 'a masterly performance' whose publication 'will form an epoch in the history of a science'. In an appropriate analogy, the weekly *Literary Gazette* praised the book for 'removing much of the delta which has obstructed the current of geological knowledge'. The *New Monthly Magazine*, despite serious reservations based on Scripture, noted that 'Professor Lyell's work is the nearest approach towards establishing geology as a science, of any thing we have met with.'[33]

No one adopted the *Principles* wholesale. It would be misleading, however, to judge the effect of any book by the number of converts or the comments of a few specialists. As with most widely read books, the effects were subtle and pervasive. Notably, critics of all shades of opinion agreed that debate about the earth would be pursued on a new plane of sophistication as a result of the *Principles*. Above all, the controversial science of geology could claim to be a worthy pursuit of gentlemen.

Freeing the Science from Moses

Geology could not be given principles without redefining the relations between human observation and divine revelation, between religious and secular education, between Church and State. In arguing these points Lyell saw no inherent conflict between science and theology; rather, he wished to redefine their respective domains. What was to be the relation between a new subject, largely imported from suspect continental sources, and native traditions of text-based theological criticism about earth history? The authority of Scripture among the learned, Lyell thought, had its foundations in popular culture (or as he condescendingly said, among the 'vulgar'). Sermons, children's books, and chronological charts on schoolroom walls, metropolitan shows, engravings of visionary paintings: the Bible structured the narrative history of the world in everyday experience. Working people were usually taught to read from the first verses of Genesis in the King James translation and assumed the literal truth of the Creation story. One boy, the son of a poor carpenter, recalled that in the 1820s no one at his school in London's East End 'could reasonably accept so doubtful an interpretation of a Biblical declaration'.[34]

Lyell, eager for the dissemination of what he saw as the truth, lamented to American friends that Britain was a country 'more parson-ridden than any in Europe except Spain'.[35] Thousands of clergymen preached Genesis as a matter of course, while no more than a handful (and here he included Sedgwick and Buckland) professed scientific views of the Flood or the earth's antiquity. Lyell realized that exegetical theology had a far longer and more distinguished history than did geology, an upstart science associated with infidelity and revolutionary atheism. For most readers, the authority of Scripture continued to outweigh that of strata maps and sections, so that biblically oriented accounts of earth history predominated in publishers' lists right through the first half of the century. Sharon Turner's *Sacred History of the World* of 1832, 'firmly attached to the great Newtonian principle, of the Divine causation of all things', went into its eighth edition two years before Lyell's book did.[36] Books in the same tradition were written by respected authors whose writings often sold more copies and were better known than those of Lyell and his friends.

No line could distinguish theological works from geological ones: Lyell's passionate desire to 'free the science from Moses' was by no means empty or non-controversial.[37] His teacher Buckland had made the most sophisticated dovetailing of scientific findings with the literary evidence of the Bible in his *Reliquiae Diluvianae* of 1823. Buckland based his defence of the physical reality of a universal Deluge on impeccable reasoning from existing causes, with texts and theological exegesis accorded a subsidiary role. He observed the actual behaviour of live hyenas in the zoo to reconstruct cave environments and ancient forms of life, and, like most practising men of science, employed a timescale far beyond the 6,000-year span accorded by tradition. There was good evidence for the Flood, too, in the unsorted debris or 'diluvium' spread across Europe. Buckland's brilliant strategy had the support of many in the moderate Church establishment, including most of the clerical geologists from Oxford and Cambridge.[38] It is not surprising that Lyell considered that the clergy were the chief obstacle to a philosophical science of the earth. Nowhere was reform more overdue than in the politics of knowledge.

The Geological Society witnessed lively debates about the Flood in the 1820s, clashes as vigorous as those taking place in Parliament over electoral reform. Together with the retired soldier Roderick Murchison and the political economist George Poulett Scrope, Lyell emerged as leader of the opposition to the diluvialist party. The *Principles,* as a manifesto for fundamental change in the organization of intellectual life, capped their campaign to sever all links of geology to a theology based on Scripture. Throughout his life Lyell continued to advocate an overhaul of the ancient universities, the centre of clerical authority; and he became a great admirer of the American republic. He saw the future of science in the hands of gentlemen of character and independence, such as Herschel, Murchison, Scrope, and himself who would draw on the best continental work without succumbing to 'unmanly' materialism or dreamy idealism.

The boundary drawn around the science in the *Principles* implicitly put clerical geologists such as Buckland and Sedgwick with those who had retarded progress. However, the polemical history of geology in the book's early chapters tactfully ends just before they had begun their work, leaving readers to draw their own conclusions. 'I conceived the idea five or six years ago,' Lyell told his ally Scrope, 'that if ever the Mosaic geology could be set down without giving offence, it would be in an historical sketch.' Lyell had to tread softly. The *Principles* could appeal to its audience only by respecting 'the weakness of human nature' and 'the sentiments of our neighbours in the ordinary concerns of the world and its customs'.[39]

This was nowhere truer than for the issue of the age of the earth. At the Geological Society, all parties in the diluvial debates had spoken freely of millions of years, but Lyell believed that they had failed to comprehend the meaning of such a vastly expanded timescale. The classic site for demonstrating the antiquity of the world was Etna, the largest volcano in Europe, and the *Principles* used it to exhibit the virtues of fieldwork as the basis of the new philosophy. The highly wrought depictions of grand scenery in the *Principles* created the imaginative depth of past time which was at the heart of Lyell's message. At this crucial moment, the narrative voice in the book became explicitly autobiographical:

> In the course of my tour I had been frequently led to reflect on the precept of Descartes, 'that a philosopher should once in his life doubt every thing he had been taught'; but I still retained so much faith in my early geological creed as to feel the most lively surprise, on visiting . . . parts of the Val di Noto, at beholding a limestone of enormous thickness filled with recent shells . . . resting on marl in which shells of Mediterranean species were imbedded in a high state of preservation. All idea of attaching a high antiquity to a regularly stratified limestone, in which the casts and impressions of shells alone were discernible, vanished at once from my mind.[40]

This was a Cartesian moment, but one especially suited to the man of knowledge in action. The philosopher did not sit in a stove-heated room as Descartes had done two centuries before, but instead explored bandit-infested country with hammer in hand. Conversion was prompted not by abstract contemplation, but by limestones, shells, and marls. And behind the reference to Descartes was a biblical one: like Saint Paul, the author changed his 'creed' on the road. The most controversial claims of the *Principles* concerned what John McPhee has called 'deep time',[41] in suggesting that other practising geologists, having abandoned the biblical timescale, had failed to realize the interpretive power of the ages at their disposal. Through the force of rhetoric, deep time was vouchsafed as a revelation from nature, realized through a secular pilgrimage.

In underlining the immensity of past time the *Principles* needed to distance its arguments from claims that the earth was eternal, for these had been commonplace in Enlightenment diatribes against religion. The issue of eternalism arose most acutely in relation to the crystalline rocks often found at the heart of mountain chains. Following the writings of Hutton and Playfair, the *Principles* argued that these rocks—previously called 'Primary'—were not remnants of an original molten state of the planet, but could in fact be formed at any time. Those that had a stratified appearance, a chapter in the third volume suggested, should be termed 'metamorphic' to emphasize that they were altered strata of diverse ages. This meant (as Lyell wrote, misquoting Hutton as usual) that 'I can find no traces of a beginning, no prospect of an end'.[42] To suggest that science should abandon the search for origins did not imply that there never was a beginning, or that there would never be an end—questions 'worthy a theologian'.[43] This reaffirmed that the *Principles* was not a cosmological treatise, but a manifesto for pursuing science. Rude attacks upon cherished beliefs before the time was ripe, as Lyell reiterated in private letters, were unjustified, unproductive, and ungentlemanly. He contrasted his strategy with that of the freethinking Edinburgh physician George Hoggart Toulmin, whose *Eternity of the World* of 1785 had been reissued by a radical publisher in the 1820s. In a letter, Lyell compared Toulmin with English soldiers who had raided Burmese temples for war trophies:

> To insult their idols was an act of Christian intolerance, and, until we can convert them, should be penal. If a philosopher commits a similar act of intolerance by insulting the idols of an European mob (the popular prejudices of the day), the vengeance of the more intolerant herd of the ignorant will overtake him, and he may have less reason to complain of his punishment than of its undue severity.

As Lyell concluded, 'This is not courage or manliness in the cause of Truth, nor does it promote its progress.'[44]

The dangers of 'insulting the idols of a European mob' were illustrated just before the *Principles* reached the booksellers. One of Lyell's closest associates, the Rev. Henry Milman, introduced German-language theological scholarship into Britain in his *History of the Jews* (1830), which did for Old Testament studies more generally what the *Principles* aimed to accomplish for the early chapters of Genesis. Milman treated the Jews as an 'eastern tribe', used documentary evidence, and minimized the miraculous. Denounced by theologians from the evangelical Sharon Turner to the High Church John Henry Newman, Milman seemed to have forfeited all chance of clerical preferment—even though it was said that nothing he had written went beyond what could be found in the notes to an expensive Bible edited by one of his most learned opponents. Milman's work, however, had been published by Murray from the first at a cheap price in the Family Library. 'The crime,' as Lyell recognized, 'is to have put it forth in a popular book.'[45]

Neither Murray nor Lyell wanted to make this mistake. Only *after* the first two editions of *Principles* had been safely launched was the set repackaged for a wider market, in accordance with Lyell's own views on the 'communication' or popularization of knowledge by gentlemen. There was intense competition in this market. The theological writer Maria Hack's *Geological Sketches* was selling well, as was an anonymous *Conversations on Geology*. Mary Shelley was impressed enough by the vogue for geology to propose to Murray an accessible work of her own. In 1834 Charles Knight published Henry De la Beche's *Researches in Theoretical Geology*, a work which argued directly against some of the ideas in *Principles*. Lyell began to worry lest the expense of his three-volume set would 'limit the circulation & my fame thereby as well as profit & give De la Beche & such compilers an advantage & they can fatten on my hard labour'.[46]

Murray recognized the affinities of Lyell's project with his own ambitions to widen the readership for his books and to beat the useful knowledge publishers at their own game. 'There are very few authors, or ever have been,' Lyell reported him as saying, 'who could write profound science and make a book readable.'[47] Lyell had originally planned his book as a series of didactic 'conversations' of the kind popular in the 1820s, and after the success of *Principles* told Murray he was contemplating a more elementary volume of dialogues at a low price.[48] Instead, the two men agreed on a third edition of *Principles* in four small volumes priced at six shillings apiece, which cut the cost of the whole work by nearly half. These were modelled directly on Somerville's *Connexion* and Marcet's *Conversations on Chemistry*.[49] *Principles* could then compete more effectively and push hack compilers out of the market. This strategy accorded with Lyell's own financial demands—he needed big sales to maintain his gentlemanly style of life—and with his lofty conception of the man of science as arbiter of truth. Through this collaboration between publisher and author, the *Principles* successfully recast for a genteel audience arguments about Scripture that in another form might have been associated with radical materialists or French revolutionaries.

This concern with respectability permeated the *Principles*. Divisive issues were introduced in measured prose, with quotations from Horace, Ovid, Pindar, Pliny, Virgil, Thucydides, Dante, Milton, and Shakespeare. The popular travel author Captain Basil Hall praised 'the calm, dispassionate, gentlemanlike style' in which Lyell handled, 'not one, but every controversial subject'.[50] The Temple of Serapis, featured as the frontispiece to the first volume (Fig. 16), reminded readers that classical structures could survive even dramatic changes. The Temple had been subjected successively to subsidence and elevation, all within the Christian era. But these upheavals, however sudden, had left the pillars standing. In human affairs as in earth history, stability was critically important.

Lyell thought that his step-by-step approach would ensure that his book 'will be thought quite orthodox' and would '*only* offend the ultras'.[51] In this he proved correct. William Wordsworth's friend

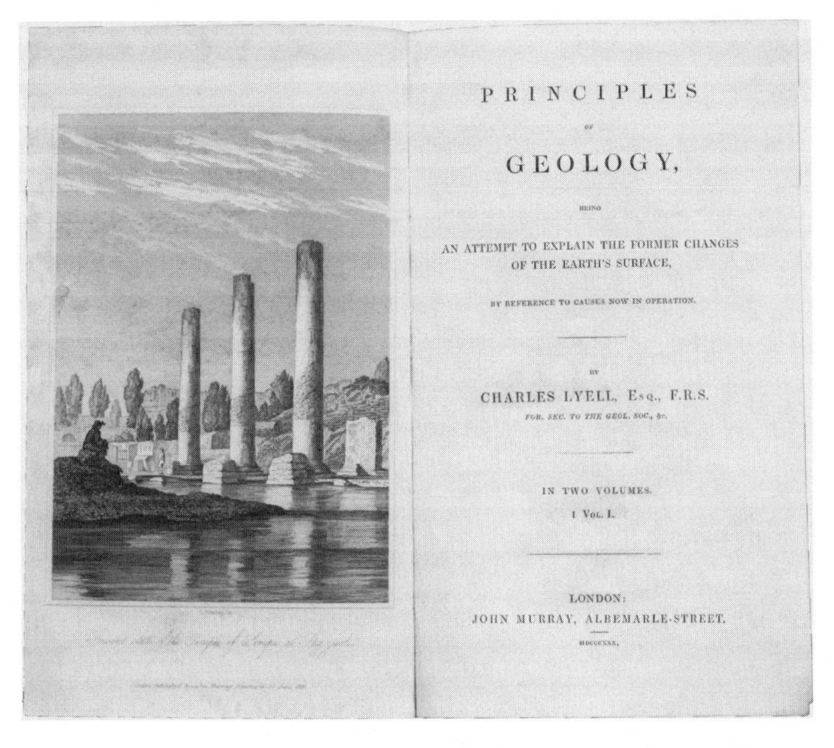

PRINCIPLES
OF
GEOLOGY,
BEING
AN ATTEMPT TO EXPLAIN THE FORMER CHANGES
OF THE EARTH'S SURFACE,
BY REFERENCE TO CAUSES NOW IN OPERATION.
BY
CHARLES LYELL, Esq., F.R.S.
FOR. SEC. TO THE GEOL. SOC., &c.
IN TWO VOLUMES.
VOL. I.
LONDON:
JOHN MURRAY, ALBEMARLE-STREET.
MDCCCXXX.

Fig. 16. The Temple of Serapis, shown as the frontispiece to the *Principles*, was simultaneously an icon of classical stability and a means to help readers understand the continuous nature of change in nature. Charles Lyell, *The Principles of Geology* (London: John Murray, 1830).

Henry Crabb Robinson, a liberal Unitarian who heard Lyell lecture during his brief tenure at King's, praised his sensitivity:

> He decorously and boldly maintained the propriety of pursuing the study without any reference to the Scriptures; and dexterously obviated the objection to the doctrine of the eternity of the world being hostile to the idea of a God, by remarking that the idea of a world which carries in itself the seeds of its own destruction is not that of the work of an all-wise and powerful Being. And geology suggests as little the idea of an end as of a beginning to the world.[52]

The religious periodical press, which blossomed in the late 1820s and early 1830s, was equally enthusiastic. As the monthly *Metropolitan*

Magazine told its readers, 'All those who wish to lift up their souls in adoration to the Great First Cause, should attentively study this work.'[53]

Even journals that supported the supreme authority of Scripture acknowledged the virtues of Lyell's book. The High Church *British Critic,* in its survey of the completed work, was almost alone among the quarterlies in confronting the *Principles* with Old Testament miracles, a 6,000-year-old earth, and a universal Deluge. However, the reviewer had rarely met with 'a more philosophical, patiently inductive chain of reasoning'. The *New Monthly Magazine* thought the author of the *Principles* 'occasionally betrayed a wildness of speculation which sets him in needless hostility to the cosmogony of Moses, without advancing his claims as a philosopher'. Even so, the reviewer praised him for writing the best book on geology ever published. The *Presbyterian Review and Religious Journal,* a new quarterly representing Scottish evangelicalism, acknowledged that the book 'is indeed a work of no common interest', 'well executed' and 'very valuable'. At the same time, the reviewer drew very different conclusions from Lyell's history of 'guessing and erring' in geological theory: 'are they all wrong but the author?' The English Nonconformist *Eclectic Review* found the *Principles*—despite its unwarranted speculations—'invaluable as a collection and arrangement of facts and geological phenomena'.[54]

Lyell achieved his greatest theological triumph within the Geological Society. The *Principles* precipitated a series of declarations from the president's chair, as former opponents abjured their use of the Flood as a middle way between Genesis and geology. As Sedgwick admitted, the 'double testimony' of the Bible and of physical traces of diluvial action had led to 'erroneous induction'. 'Having myself been a believer,' he told the Society, 'and, to the best of my power, a propagator of what I now regard as a philosophic heresy, and having more than once been quoted for opinions I do not now maintain, I think it right, as one of my last acts before I quit this Chair, thus publicly to read my recantation.'[55] The *Principles* proved decisive in the boundary disputes between science and theology.

The Dangers of Transmutation

If the Flood and the age of the earth ceased to be questions of debate at the Geological Society, they occupied the centre stage of public concern for another three decades. (In many parts of the world, including the United States, they remain controversial to the present day.) Such discussions remained urgent partly because atheists and deists, writing in the working-class penny press, used geology to give the stamp of authority to unbelief. But the infidels had hard work to make a book whose last sentence compared 'the finite powers of man and the attributes of an Infinite and Eternal Being' into an ally of atheism.[56] The *Principles,* in offering new proofs of divine power and goodness, had distanced geology from Scriptural exegesis, miracles, and Creation, but not from all theology. At the same time, though, it cut away the supports that science had traditionally provided for belief. In the *Principles* the rock of faith rested solely on God's maintenance of the economy of nature and on the status of humans as moral, rational beings.

In stressing the special character of the human species, the *Principles* drew on the Bishop of Chester, John Bird Sumner's *Records of Creation,* which used the distinction between humans and animals as evidence for a divine creator. To claim otherwise had been the mistake of the French materialists, most notoriously Jean-Baptiste Lamarck. In his *Philosophie zoologique* of 1809 and the introductory essay to his definitive seven-volume work on invertebrates, Lamarck had argued for the evolution of one species into another, or 'transmutation' as it was generally called, within the context of a broader philosophy of nature which emphasized the power of environmental circumstance.

An attack on transmutation became central to the *Principles* because of the threat Lamarck posed to the special status of humanity.

During visits to Paris, Lyell had been shocked to discover that transmutation 'has met with some degree of favour from many naturalists',[57] among them Étienne Serres, Étienne Geoffroy Saint-Hilaire, Jean-Baptiste Bory de Saint-Vincent, Pierre Boitard, and Jean-Baptiste d'Omalius d'Halloy. In Britain, Robert Jameson's lively scientific monthly, the *Edinburgh New Philosophical Journal,* had published an anonymous article advocating transmutation. Perhaps most threateningly, the declared Lamarckian Robert Edmond Grant was on the Geological Society Council and taught fossil zoology at London University.[58]

But Lyell had few worries about these views spreading among his intended readership, where transmutation was typically dismissed as 'the most stupid and ridiculous' idea to have been hatched by 'the heated fancy of man'.[59] The second volume of the *Principles* began by noting that Lamarck's views were 'not generally received'.[60] Even Lyell had had to read a borrowed copy of the *Philosophie zoologique,* which was not translated into English until 1914. By refuting the French naturalist at length, the *Principles* made transmutation accessible throughout the English-speaking world, beyond a narrow circle of naturalists, medical lecturers, and political radicals. The real question is not why the *Principles* should reject Lamarck, but why it did so publicly and at such length. One answer becomes evident if the audience for the *Principles* is kept in mind. Lyell did not expect his book to be read by atheists or deists, but by conservative, genteel subscribers to the *Quarterly*. This audience needed to be shown that geology was safe. Lyell anticipated clerical attacks on his advocacy of the uniformity of nature, views which could lead to suspicions that the groundwork was being laid for a naturalistic explanation of species. However minor an issue transmutation might be within science, annihilating Lamarck established Lyell's orthodox credentials.

Given the prestige of French science and its importance in his attempted reform of geology, Lyell's fear that he might be condemned as a closet Lamarckian becomes understandable. If uniform causal laws applied to all other parts of the natural world, why not to humanity itself?

The need to combat transmutation more directly had become clear to Lyell only at a very late stage in the composition of his first volume. One factor was certainly the impending publication, by Murray, of Davy's posthumous *Consolations in Travel,* which linked the 'absurd, vague, atheistical doctrine' of transmutation with Huttonian views on uniformity.[61] Lyell must have been shown the relevant passages in proof and realized that views similar to his own were suspected of tending to materialism. In passages added at the last minute, the first volume of the *Principles* countered the growing alarm. These did not explicitly contest the claim that uniform natural laws entailed species transmutation, but undercut the argument at a more fundamental level by denying the assumption of a progressive history of life.[62] No progress, no evolution: the method of uniformity, far from leading to the *Philosophie Zoologique,* made the views in that book impossible.

Lyell strategically refrained from mentioning the evolutionary spectre at this stage of his book, despite abundant opportunities to do so. His increasing awareness of the dangers of being tarred with a Lamarckian brush, however, soon led to major changes in his second volume. His original plan, submitted to Murray in 1827, had devoted no more than four or five chapters to changes in the organic world with (at most) a brief dismissal of Lamarck. At a later stage—probably when he returned to his old drafts in November 1830, when *Consolations* had been out for nine months—these chapters were greatly expanded, with the attack on transmutation extended well beyond its allotted space. The new chapters cited evidence from animal and plant hybrids, the similarity of mummified animals from ancient

Egypt to present-day forms, and the variability of domestic forms, to demonstrate that there was no evidence for transmutation operating as a *vera causa*.

Reactions to the first volume, especially by Sedgwick in a Geological Society presidential address, established the virtues of the new strategy. For Sedgwick, as for Davy in *Consolations*, the thoroughgoing application of 'what we commonly understand by the laws of nature' to the problem of life would seem to demand 'the doctrines of spontaneous generation and transmutation of species, with all their train of monstrous consequences'.[63] As Lyell recognized, distancing his writing from Lamarck was just as imperative as liberating science from Scripture.

Lyell also confronted career pressures on the issue of transmutation early in 1831. As part of his campaign to establish himself in the metropolis, he was negotiating for the geological chair at King's College. One of the electors was Edward Copleston, who by this time was Bishop of Llandaff and Dean of St Paul's. Copleston feared that geology undermined beliefs concerning the reality of a universal Deluge, and the divine creation of new species, especially man. Lyell could satisfy him in part on the first point, although he equivocated by stating that although the Flood had not been universal, it might have covered the inhabited globe.[64]

On species Lyell had no reservations; 'I combat in several chapters of my second volume the leading hypothesis which has been started to dispense with the direct intervention of the First Cause in the creation of species, & I certainly am not acquainted with any physical evidence by which a geologist could shake the opinion generally entertained of the creation of man within the period generally assigned.'[65] Sedgwick and Conybeare intervened on Lyell's behalf with the electors at King's, aware that noxious Lamarckian dogmas were dispatched in the new chapters. Cannily and cautiously, Lyell never said in public if a law-like origin for species was possible, preferring his views 'to be inferred'. In correspondence he tailored his message to the recipient. To liberal friends like Herschel, he advocated 'the intervention of intermediate causes'; to vigilant clerics like Copleston, he appeared to confirm 'the direct intervention of the First Cause'. When cornered, Lyell could allay orthodox suspicions by arguing that origins were not part of science.[66]

The anti-Lamarckian additions were sometimes recognized as tangential to the main argument of the *Principles.* 'Lyell is amusing,' the Edinburgh naturalist Robert Jameson wrote, 'but it is not what geologists want.'[67] The *Gentleman's Magazine* was disappointed too, after being 'compelled to wade through one half of the volume among *disjecta membra* of the vegetable and animal kingdoms, which might do, as speculations on natural history, or under any other title, than that of "Principles of Geology"'. A handful of reviewers complained that Lamarck was an outdated straw man and that the *Principles* would have been more convincing on this issue had it targeted a more credible opponent, such as the transcendental anatomist Étienne Geoffroy Saint-Hilaire. The *Literary Gazette* pointed out that the real debate centred on continental discoveries about monstrosities and embryology—not on crude jokes about savages and orang-utans. 'It is always much more easy to ridicule a theory,' the reviewer noted, 'than to prove its soundness.' Even the *Spectator,* for all its enthusiasm, thought that Lyell had been successful only in undermining Lamarck's evidence—not transmutation itself.[68] But most critics, including those in the *Quarterly,* felt that the *Principles* had successfully parried Lamarck. With its elegant rhetoric, strategic silences, and wealth of evidence, the discussion was praised as a model of clear argument.

Yet there was another side to Lyell's response, as revealed in his private notebooks on species. These extraordinary documents reveal an author appalled by the religious, political, and ethical consequences of transmutation. If transmutation was true, these notebook entries suggested, no divinely implanted reason, spirit, or soul would set human beings apart; they would be nothing but an improved form of the apes that he watched, fascinated, at the newly opened London Zoo. Transmutation was a dirty, disgusting doctrine, which raised fears of miscegenation and sexual corruption. Not only did transmutation repel Lyell's highly refined aesthetic sense, it undermined his lofty conception of science as the search for laws governing a perfectly adapted divine creation. With humans no more than better beasts and religion exposed as a fable, the foundations of civil society would crumble, just as they had done in revolutionary France. As Lyell admitted to Darwin many years later, it was Lamarck's conclusion 'about man that fortified me thirty years ago against the great impression which his arguments at first made upon my mind'.[69] These secret fears shaped not only the explicit attacks on transmutation, but the entire geological system of the *Principles*.

Lyell read the *Philosophie zoologique* in 1827, and in a letter to Mantell compared the experience with the pleasures of light fiction. 'His theories delighted me more than any novel I ever read,' he wrote, 'and much in the same way, for they address themselves to the imagination, at least of geologists who know the mighty inferences which would be deducible were they established by observations.'[70] This has sometimes been seen as a positive reaction, but in the light of contemporary scepticism about novel reading it was about as damning a comment as a man of science could make. Lamarck addressed the imagination rather than the reason and relied on invention rather than observation. Lyell admired 'even his flights' but thought him wrong.

The shock of reading Lamarck led Lyell to take up some of his most controversial positions in the *Principles*. It was for this reason that the work denied Buckland's grand narrative of progress, for progress implied a regular gradation of organisms in time, with all that meant for the status of human reason and the soul. Instead, Lyell stressed the geographical causes of climate change and the piecemeal nature of the geological record. These, as we have seen, became key tenets of the *Principles.* Confronting Lamarck also led Lyell to put the human species outside the flux of natural causation, a position many thought was strangely inconsistent with his views on natural law. Even the humble molluscan fossils of Etna were pressed into his anti-evolutionary crusade. As he wrote from Paris in 1830 when revolutionary crowds once again thronged the streets, if species of shells remained unchanged after thousands of years, it 'must therefore have required a good time for Ourang-Outangs to become men on Lamarckian principles'.[71]

This distaste for bestial origins remained deeply private, a matter of personal faith inappropriate for airing in the public world of quarterly reviews, scientific meetings, or London clubs. Lyell thus concealed his fears behind a public stance that all recognized as scientific. On such grounds, the *Principles* could mount a credible case against the widespread assumption of a progressive history of life and reject Lamarckian transmutation as an imaginative fantasy. Contemporary readers, even Lyell's closest friends, never learned of the pervasive significance of the status of the human species for his science.

The Trojan Horse

For the next fifty years the *Principles* held its place as a standard work, one of the key titles on Murray's list. Numerous editions appeared in the United States, but the *Principles* made less of an impact in continental Europe, presumably because of the very different theological and political setting—although French and German translations were eventually issued.[72] Its wealth of detail from the Andes to the Malay Archipelago encouraged readers to adopt a global perspective on processes of geological change. Among the next generation of naturalists, the *Principles* established frames of reference that took scientific writing in new directions. For naturalists and explorers the book provided new ways of incorporating unpredictable exotic landscapes within a framework of scientific explanation. Darwin became Lyell's most enthusiastic advocate, and found the *Principles* immensely liberating from the moment of landing on his first tropical island during the *Beagle* voyage. The young naturalist used the book to develop his own causally oriented style of interpretation, comparing the uplifted but scarcely disrupted rocks in the mid-Atlantic Cape Verde Islands with the Temple of Serapis. Later in the voyage he reinterpreted the origin of coral reefs, overturning the model that Lyell had advocated; but the style of reasoning was that of the master, who altered his next edition accordingly.[73] When Darwin began to consider the possibility of species transmutation after his return home, he did so in private dialogue with the *Principles*, little suspecting that it was the author's fears of such views that had shaped the book.

In the context of reform, the *Principles* was a Trojan horse. It had the classically balanced prose, learned references, and gentlemanly form of publication to shift the terms of debate among those whom Lyell recognized as the holders of power. Opposing divine intervention in nature, combating the Flood, and hammering home the consensus of geologists on the antiquity of the world, the *Principles* could have been seen as allied to Enlightenment freethought, and suffered a fate like that of Milman's *History of the Jews*. But the reaction was overwhelmingly positive. The cautious rhetoric, the clever demolition of Lamarck, and the sheer empirical exuberance of the *Principles* made it possible to revitalize a lawful system based on actual causes, and to exorcize the demons of revolutionary deism which had dogged the study of the earth since the 1790s. Only someone with all the independence of character and sense of propriety that being a gentleman entailed could have used such an unlikely vehicle to advocate controversial views. The *Principles* had the imprint of conservative classicism, but hid within a secret army of reform.

6

The Problem of Mind

George Combe's *Constitution of Man*

> Mr Chambers was requested a few days ago to show his steam-printing press to a party of ladies, and one of them took up a sheet that had just issued from it; when she saw that it was the 'Constitution of Man' she threw it down as if it had been a serpent, and exclaimed—'Oh, Mr Chambers, how *can* you print that abominable book?'
>
> —George Combe, 16 July 1836

The steam-powered printing machine run by the publishers William and Robert Chambers was a spectacle of the industrial age, well worth a visit to their premises just off the High Street in Edinburgh's Old Town. Newspapers such as *The Times* had been printed by machine for over two decades, but the Chambers brothers were applying these innovative technologies to the literature of self-improvement and reform. On this occasion the sheets flying off the press were from *The Constitution of Man, Considered in Relation to External Objects,* which had been published for a relatively limited readership in 1828 but was now the first in a series of extraordinarily inexpensive 'People's Editions'. The author of the *Constitution,* the Edinburgh lawyer and lecturer George Combe, saw his book as a revolutionary attempt to bring an understanding of human action into the realm of law, and as the founding document in a campaign to create a science that could ensure the future of the human race. As the shock of the visitor to the Chambers's printing factory suggests, however, the *Constitution of Man* could also be seen as advocating views about God, man, and nature that could be condemned as 'abominable'. The very idea that the mind might depend on the brain's physical qualities was extremely controversial, and none of the works discussed so far were willing to advocate it. Lyell's *Principles of Geology* characteristically presented reason as a divine gift that ensured mankind's special status and kept the moral history of humanity separate from the physical history of the earth.

At the heart of reformulating scientific knowledge in such works was the question of the human mind. Was there something unique about human rationality and cognition, and if so what was it? What was the relation between an individual's moral, spiritual, and intellectual endowment and circumstances of upbringing and environment? These were questions that had been asked over thousands of years, but many readers expected they could now be tackled with the methods and instruments of science. A wide range of scientific approaches to mental phenomena were developed in the decades after the 1790s, drawing on late Enlightenment projects which applied the categories of reason to humans. New mental states, potentially involving psychic influence at a distance through mesmerism or spiritualism, opened up exciting possibilities. The old dichotomies between body and mind were being abandoned, with philosophical methods giving way to psychological and ultimately to physiological ones.

For millions of readers throughout the world, this great task was accomplished by Combe's *Constitution of Man*. As with many of the new sciences of this period, important aspects of the views expressed in *Constitution* were continental imports. Initially proposed by Franz Joseph Gall around 1800, the science that became known as 'phrenology' was effectively the creation of the German physician Johann Spurzheim, who lectured throughout Europe and America. The

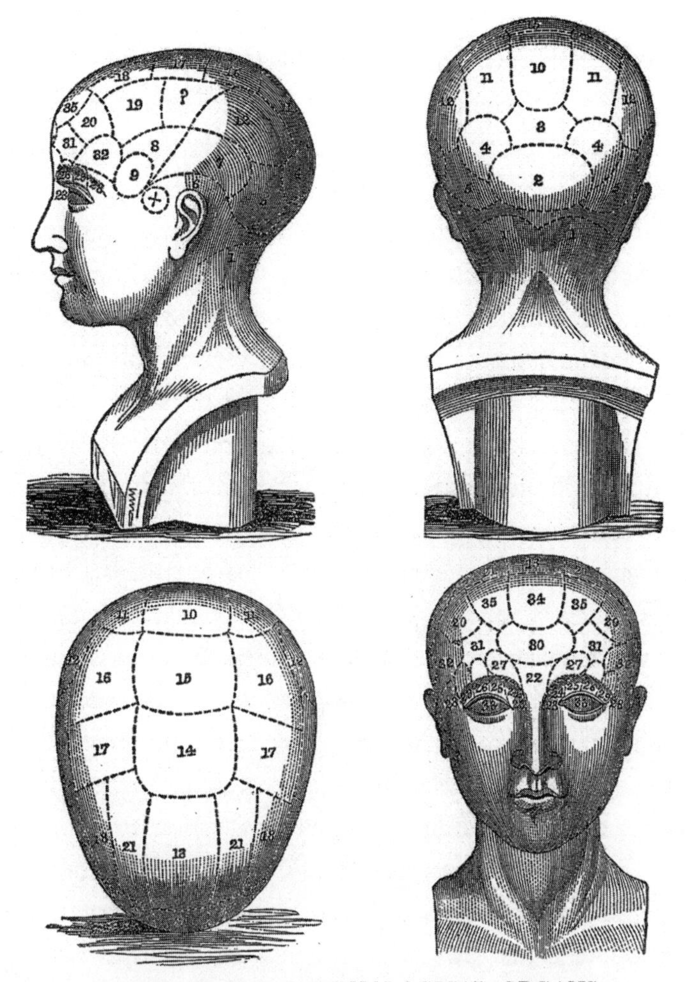

NAMES OF THE PHRENOLOGICAL ORGANS,

REFERRING TO THE FIGURES INDICATING THEIR RELATIVE POSITION.

AFFECTIVE.		INTELLECTUAL.	
I. PROPENSITIES.	II. SENTIMENTS.	I. PERCEPTIVE.	II. REFLECTIVE.
1 Amativeness P. 8	10 Self-Esteem . P. 12	22 Individuality P. 18	34 Comparison P. 23
2 Philoprogenitive-ness, ... 8	11 Love of Approbation 12	23 Form . . . 19	35 Causality . . 24
3 Concentrativeness 8	12 Cautiousness . . 13	24 Size 19	
4 Adhesiveness . 9	13 Benevolence . . 13	25 Weight . . . 19	Modes of Activity 24
5 Combativeness . 9	14 Veneration . . . 14	26 Colouring . . 19	Practical Directions 28
6 Destructiveness 9	15 Firmness 15	27 Locality . . 20	Combinations . 29
+ Alimentiveness 10	16 Conscientiousness 15	28 Number . . 20	Materialism . . 31
7 Secretiveness . . 10	17 Hope 16	29 Order . . . 20	
8 Acquisitiveness 11	18 Wonder 16	30 Eventuality . 21	
9 Constructiveness 11	19 Ideality 16	31 Time . . . 21	
	? Unascertained.	32 Tune . . . 21	
	20 Wit or Mirthfulness 17	33 Language . . 22	
	21 Imitation . . . 17		

Fig. 17. 'Names of the Phrenological Organs', from George Combe, *Outlines of Phrenology* (Edinburgh: Maclachlan, 1836), frontispiece.

crucial move in phrenology was to make the brain the organ of the mind. Like other parts of the body, different parts of the brain could then be seen as 'organs' specializing in different functions, from physical love to secrecy, from mathematical calculation to religious devotion (Fig. 17). The size of different parts of the brain reflected their relative development in each individual, and could be interpreted (by those who knew the science) from the external shape of the skull. Understanding the qualities of the brain in this way meant that hidden qualities of character could be 'read' by an informed observer. (Leger's magnetoscope, as used to analyse the brain of Charles Babbage, was obviously a different means towards the same end.) A true phrenological reading required a shaved head: under the fingers of a trained phrenologist, the skull would not lie.[1]

Constitution stressed that possessing a clearly articulated theory of mind was essential for living according to the natural laws, as the first step towards self-help and not being misled by others. The natural laws were divided into three kinds: physical, organic, and moral and intellectual. Physical laws, like gravity or chemical attraction, applied to all matter. Organic laws involved questions of generation, growth, and decay, and were limited in their application to living beings. Intellectual and moral laws related only to conscious beings. Within this latter category, animals such as dogs and horses could be seen as subject to intellectual laws, while humans acted on the basis of both intellectual and moral laws. Combe held that each category of law was entirely distinct, but that all needed to be understood and obeyed to avoid catastrophe. A shipload of missionaries might obey every letter of the moral law, but if they failed to appreciate the physical law governing the tides, they would drown. For the first time, readers could achieve the self-knowledge that would allow them to change their circumstances and act appropriately in a rapidly changing and increasingly individualistic world. *Constitution* provided a recipe for human action in everyday life, from choosing a partner to approaching the great questions of belief.

Reading Constitution

In 1835 Margracia Loudon was at a turning point in her career as an author. She had lived in the Midlands resort town of Leamington Spa from the time of her marriage five years earlier, and had written three-volume novels about fashionable life, with titles such as *First Love* and *The Dilemmas of Pride.* A considerable celebrity, she became a prolific writer in the monthly and quarterly periodical literature. At the age of forty, however, her writing had taken a new and ambitious direction with *Philanthropic Economy*, which intervened in the debates about the post-reform Whig government's attempts to tackle poverty. Together with her husband, a successful physician who had written on the local spa waters, she was active in local liberal intellectual circles, particularly those associated with Unitarianism and dissent from the Church of England. She was delighted by the reviews of *Philanthropic Economy*, and planned a sequel which would pursue further the ideas in her chapter on the 'Philosophy of Happiness'.[2]

Loudon asked a friend, James Kennedy, a doctor with a strong interest in natural history and science, for advice about what to read. Kennedy attempted to convert anyone to phrenology who cared to listen, although the response was often no better than indifference, as when a clergyman's wife 'both intelligent and well-informed', fell asleep when reading debates on the subject in the *Edinburgh Review*, and only narrowly 'escaped a fit of "hystericks" on being startled from her doze' when the blue-bound volume hit the floor.[3] Kennedy expected Loudon would be almost as difficult to convince. He had previously tried to introduce her to the science, and blamed her lack of interest on her preoccupation with writing as well as her 'dread of the bastard ridicule begotten by focussing upon falsehood, which has so plenteously poured forth upon the phantasms of "Bumps" and "Craniology"'. In response to her request, Kennedy sent Loudon suggestions for reading and the latest expanded version of *Constitution*.[4]

Loudon read the book three times, following precisely Kennedy's instructions to 'examine every sentence and every page of it with impartial and profound attention'. The work began by surveying the external world and organic life in relation to the human race. It was not an 'optimistic' system, but one of constant progressive improvement. Man, although in many ways an animal, had 'higher powers' bestowed by God, and 'the grand object of our existence' was to use these as fully as possible. After laying down the foundation of the laws of nature, the work then discussed man as a physical, organic, and moral and intellectual being, and showed how this understanding could be used in daily life, from choosing servants to achieving the highest happiness. Human misery was not due to man's fallen, sinful nature (as most orthodox Christians would argue) but to a failure to observe the laws of nature. A final chapter argued that the new science of mind was compatible with the revealed biblical truth. As Loudon reported:

> ... never was I so entirely delighted with any book. That one small volume seems to me to comprise more wisdom, of a kind practically applicable to the production of human happiness, than all the ponderous works put together that I have ever met with. All the vague aspirations after any thing true or beautiful, which I have ever traced, faintly gleaming through the mystic or inconsistent pages of other writers, appear to me to be, in Mr. Combe's book, concentrated into a steady, cheering, guiding light, by which mankind, if they would but accept the aid it offers them, might feel assured of attaining true felicity, both temporal and eternal.

Constitution, she declared, 'may possibly have given a direction to the pursuits of my whole life'.[5]

Self-confident, opinionated, and aware of the power of her own fame, Loudon presented herself as a disciple. 'When I read such a work, I feel as though it would be folly and presumption in me ever to think of writing any thing again; that I might rather to sit down, for the rest of my life, like a child at the feet of such men, learning their wisdom'. When so few men had real learning, she told Kennedy, 'unlearned ladies' had to rely on feminine common sense lest they be misled by 'dogmas full of nonsense and contradiction'. She wanted to talk with Kennedy (what she had to say 'would cover as much paper as a three volume work') and above all to meet the author of *Constitution*.[6] She collected newspaper clippings of Combe's teachings in a folder entitled 'Beautiful Lectures'; dreamed that he would come to England and stay at their house; and even though a stranger, felt compelled to correspond in the language of friendship.[7]

For readers like Loudon, encountering *Constitution* sparked an experience akin to a religious conversion, a moment of transformation inspired by the prophet's words on the page. The book was her 'great guide', which expressed 'my inmost thoughts'. While prostrating herself before the author, however, Loudon also made it clear that she recognized the value of her conversion, especially at a time when *Constitution* had yet to receive much public notice. She told Combe she could engineer access to the prestigious liberal quarterly *Westminster Review*, and although the promised review did not appear, she did put her name to a prominent notice in the Unitarian *Monthly Repository*.[8]

Like many adherents of *Constitution*, Loudon saw the future of phrenology not in the elaboration of further chapters in a system, but in plans for practical education and political reform. She promised Combe that her planned philosophical book would be based on phrenological principles, and in 1845 the work begun a decade earlier as a 'philosophy of happiness' was serialized in a literary monthly and then published as *The Light of Mental Science; Being an Essay on Moral Training*. The vital issue, the new work argued, was the effective management of the highly impressionable minds of infants and children, especially in relation to the key faculty of 'benevolence'.

This meant transforming the education of those engaged in childcare, whatever their gender or station in life:

> Now, neither the manufacturing labourer nor the carpenter's apprentice could have deduced the rules from the science; but they have each been taught to apply the rules, when so deduced, to their daily work. Why, then, should not nursery maids be taught to apply plain practical rules, deduced from mental science, to their daily work—that of *influencing* the *associations* of the infant mind?
>
> But philosophers take no cognizance of the existence of nursery-maids, and mothers only ask if they can do up small linen. Yet, if philosophers have condescended to assist the manufacturer and the carpenter, may they not aid the nursery-maid and the mother?[9]

The Light of Mental Science made ambitious suggestions for a new system of learning and teaching, advocating a national committee and a universal worldwide parliament. The result would be the end of wars, crime, and human misery.

Combe, in acknowledging a presentation copy of *Mental Science*, could not conceal his anger. What bothered him most was the book's overeager stress on the potential for individuals to improve themselves, for in his view phrenology demonstrated strict limits to the development of the faculties, even if education commenced in infancy. To think otherwise was 'a grave fallacy, unsupported by nature, disowned by experience, & calculated to lead only to disappointment'.[10] Understanding the faculties could enable a rebalancing towards virtue, but gross deficiencies could not be made up. Loudon had, by this account, gone too far in assuming limitless human improvability, of a kind associated with the utopian socialism of Robert Owen and his followers, including Combe's brother Abram, who had lost his fortune and health in a utopian scheme for communal living. *Constitution* included a moving account that drew lessons from this example:

> The individual here alluded to, was deeply engaged in a most benevolent and disinterested experiment for promoting the welfare of his fellow creatures; and superficial observers would say that this was just an example of the inscrutable decrees of Providence, which visited him with sickness, and ultimately with death, in the very midst of his most virtuous exertions. But the institutions of the Creator are wiser than the imaginations of such men. The first principle on which existence on earth, and all its advantages depend, is obedience to the physical and organic laws. The benevolent Owenite neglected these, in his zeal to obey the moral law; and, if it were possible to dispense with the one, by obeying the other, the whole theatre of man's existence would speedily become deranged, and involved in inexplicable disorder.[11]

Socialism advertised itself as a 'science of the influence of circumstances', while phrenology started from the inbuilt qualities of the brain. Many radicals had been enthusiastic about the science, but it was becoming clear that its challenge to the established order was strictly limited. Combe consistently argued that the present state of society was the result of divine laws and that people should be contented with their stations in life.[12] Loudon—although certainly no socialist—believed that the distinctive psychological faculties described in phrenology could be enhanced by association and example.

The most striking divergence between Loudon and Combe, however, was over tactics, for *Mental Science* avoided any reference to phrenology. This represented a major change of strategy on Loudon's part, for her review of *Constitution* in 1836 had stressed that reformers needed to be explicit about their debts to the new science of mind, which provided 'an argument of infinitely greater force' than vaguely gesturing to '*some reason or other*'.[13] It also reflected a more general decline in the willingness of liberals to acknowledge their often substantial private debts to phrenology and to Combe, lest they be seen as naïve, unfashionable, or irreligious. Even with this omission, however, neither *Mental Science* nor the serialized essays were successful. Loudon went on to write a novel that thinly disguised the same teachings in a story about the love of a mother for her children.

Earlier in their correspondence, Combe had diagnosed the reasons why a woman like Loudon should be so profoundly affected by reading *Constitution*. In some authors, he explained, the intellect was inspired by the moral sentiments, and in others the reverse held, with the intellect providing the starting point. In Loudon's case, the sense of morals was 'ready, prolific, warm, & energetic', but as the utopian character of her proposals in *Philanthropic Economy* demonstrated, she wanted the direction towards utility that the understanding of natural law in *Constitution* could supply: 'moral sentiments unenlightened by this philosophy appear like blind Polyphemus groping in the cave & desiring to get out, but never knowing how to escape into day'.[14] Without a correct understanding of phrenology and its consequences as expressed in *Constitution*, she was lost. Combe thought this was the great flaw of Babbage too, although his brain in most ways was so completely different: his writings 'were a curious specimen of a vigorous mind wanting a science of man's nature to render its views harmonious and sound'.[15]

Making a Science of Mind

Constitution aspired to be the necessary 'science of man's nature' hitherto lacking for readers like Loudon and Babbage. Combe was well aware from his own struggles that the science of mind was one of the most contentious domains of knowledge. As a young barrister in Edinburgh during the second decade of the nineteenth century, he had found himself unable to accept traditional Christian understandings of human sinfulness and redemption. He had read widely in the philosophy of mind, but had been frustrated in his attempts to apply the lessons of John Locke, Adam Smith, and Dugald Stewart to everyday life. His attendance at public dissections on the anatomy of the brain ended in disappointment when the lecturer John Barclay said that cerebral structure revealed nothing about the workings of the mind. Although sceptical about phrenology, in 1815 Combe attended a private lecture by Johann Spurzheim, who was touring Scotland in the wake of vicious attacks in the *Edinburgh Review*. Spurzheim presented a choice: on the one hand, the printed pages of the *Edinburgh*, on the other a dissection of the brain that (unlike Barclay's) actually seemed to illuminate the functions of mind. This was a turning point for Combe. Moral and mental phenomena had to be part of science.[16]

The science of mind Combe advocated, however, was of a particular sort. Despite being impressed by Spurzheim's performance with the scalpel, he was most engaged in developing the wider implications of phrenology for human affairs. Combe saw the subject as one that would be pursued primarily through skilled observations by experienced practitioners, rather than through dissection or physiological experiment. Because phrenological science involved the mind, he believed it could not be based on 'quantity and number':

> Phrenology is not an *exact*, but an *estimative* science. It does not resemble mathematics, or even chemistry, in which measures of weight and number can be applied to facts; but, being a branch of physiology, it, like medical science, rests on evidence which can be observed and *estimated* only. We possess no means of ascertaining, in cubic inches, or in ounces, the exact quantity of cerebral matter which each organ contains, or of computing the precise degree of energy with which each faculty is manifested; we are able only to *estimate* through the eye and the hand the one, and by means of the intellect the other.[17]

There were, of course, good reasons for pursuing this course in Edinburgh after the Burke and Hare scandal, when vagrants and prostitutes had been killed to provide bodies for dissection in university anatomy classes. More fundamentally, however, Combe believed that the knowledge required was of a different kind. Phrenology did not involve viewing the body analytically as a machine that could be taken apart to understand its internal relations. Instead, it was a science of surfaces and classification, diagnosis and visual observation in the consulting room and in museums of casts and skulls. What mattered was the skill embodied in the hand and eye of the trained phrenologist.

This meant that phrenology, like most contemporary medicine in Britain, would be pursued through the study of cases. Every individual was different, and each brain therefore the nexus of a complex web of birth and circumstance, resulting in physical characteristics which could be understood through the touch of the phrenologist proficient in 'reading' the contours of the skull. Measuring tools were important, from calipers that gauged cerebral development, to (more unusually) devices such as the magnetoscope applied to Babbage's brain, which measured emanations of magnetic force. But what really mattered was the skilled movement of the fingers of the trained phrenologist. This skill could be developed, not by memorizing a chart of the different organs, but through practice. Part of this was inevitably 'hands on', but reading the details of well-attested cases and building an understanding outwards from those could also play a vital role.

This stress on cases is clear from the publication patterns of phrenology, which were modelled on medical and natural historical ones. Combe was at the centre of a network in the 1820s and 1830s that attempted, in a heroically single-minded way, to take phrenology out of the realm of controversy by providing the full range of printed works that would mark it as a science.[18] For a subject to appear appropriately systematic and progressive, readers needed primers, introductory surveys, advanced textbooks, and journals. In the Scottish educational context, such a range of publications marked out a body of knowledge that might compete for space on the curriculum, an issue that became of particular significance when Combe ran for the chair of logic at the University of Edinburgh in 1836. More important, at a time when science of any kind was rarely taught in schools and universities, such a range of publications encouraged a process of self-education, in which readers could move securely through successive stages of understanding.

The trajectory away from polemic is apparent in the successive editions of Combe's first book. When first published in 1819, *Essays in Phrenology* was a sharply worded defence against the *Edinburgh Review* and other sceptical critics. In 1825 the work was recast as a *System of Phrenology*, which surveyed the principles of the subject and then methodically described the mental faculties, categorized by their order and genus. In this new form the book was modelled on standard works of scientific reference such as Robert Jameson's *System of Mineralogy* and James Edward Smith's *Introduction to Physiological and Systematic Botany*. In its fifth edition of 1843 the *System of Phrenology* occupied two thick volumes. A pamphlet-length *Outlines of Phrenology* and an introductory *Elements* were also available from the mid-1820s. A similar attempt to move towards respectability is evident in the history of the *Phrenological Journal*, founded in 1823 by Combe and his supporters as a forum for discussion free of the anti-phrenological biases of the Whig *Edinburgh* and the Tory *Quarterly* reviews. Under a succession of editors, the reporting of new case studies in the *Phrenological Journal* became a significant goal. The most ambitious work in this programme of publication was Combe's *Constitution*, with its explicit aim of claiming for the new science much of the territory occupied by traditional moral philosophy and revealed religion.

Of all the forms of useful knowledge in the years around 1830, a scientific understanding of the springs of human action was the great prize, and *Constitution* was only one among a host of treatises setting out a bewildering array of possibilities. Many of these were handsome octavos, building on the philosophical traditions of the Enlightenment. The Glasgow preacher Thomas Chalmers's *The Adaptation of External Nature to the Moral and Intellectual Condition of Man*, published in 1833, was one of the eight Bridgewater Treatises exploring God's relation to Creation in light of the latest scientific findings. In an uninvited addition to the Bridgewater series, Babbage demonstrated the potential of his calculating machine to reproduce aspects of human intelligence. In moments snatched from his duties as Lord Chancellor, Henry Brougham wrote on instinct, and the ageing radical William Godwin compiled *Thoughts on Man,* which argued that individuals should be allowed to pursue their abilities to the farthest extent possible, whatever their social status. Most of these works ignored one another and had no time at all for phrenology, which Godwin ranked with 'chiromancy, augury, astrology, and the rest of those schemes for discovering the future and unknown, which the restlessness and anxiety of the human mind have invented, built upon arbitrary principles'.[19] Godwin, by this time isolated from mainstream intellectual debate, was almost equally sceptical about the findings of Newtonian astronomy, and his book was largely ignored too.

The great fear for many readers—although there were many ways of resolving the problem—was that a scientific understanding of the mind would conflict with Scripture. As in the case of geology, the most significant issue was the special status of humans, and the great bugbear was 'materialism'. Radicals such as Richard Carlile reinforced this notion by reprinting classic Enlightenment works such as those of the revolutionary American deist Elihu Palmer and the populist democrat Tom Paine. Respectable publishers, when unavoidably committed to accepting manuscripts that veered dangerously close to such doctrines, limited sales with tiny print runs and extortionate prices. Against his better judgment, in 1831 Murray published the neoclassical furniture designer Thomas Hope's three-volume *Essay on the Origin and Prospects of Man,* which advocated the transmutation of species as part of a system of abstract metaphysics. Released in an expensive edition of only 250 copies, the work was almost impossible to obtain. In the same year the antiquarian publisher John Bowyer Nichols prepared to issue a *Psychology* written anonymously by the ultra-Tory opponent of reform, the Cornish baronet Sir Richard

Vyvyan, but its arguments for the immortality of the animal soul proved so explosive that publication had to be stopped.[20]

Materialist philosophies of mind had long been available to the wealthy and discreet, and were sold in much the same way (and sometimes through the same channels) as pornography. The problem by the 1820s was that such theories were also available for a few pennies to much larger and diverse readerships. As a costly three-volume treatise, Hope's *Essay* was virtually inaccessible, especially given its vast sentences (sometimes thousands of words before reaching the verb), but skilfully extracted in the weekly *Literary Gazette* it was available at least to a middle-class audience. The most notorious accusation of infidel materialism centred upon the surgeon William Lawrence's *Lectures on Comparative Anatomy, Physiology, Zoology, and the Natural History of Man,* first published in 1819 as a student text by the medical publisher John Callow. Lawrence had argued that the human mind was the product of man's superior organization, so that a bloody dissecting room was no place to search for the human soul. When the Lord Chancellor declared these views blasphemous in an Old Bailey trial, the book lost the protection of copyright and was immediately reprinted by Carlile and other radicals at bargain prices.[21] By the early 1830s, the atheist publisher James Watson had extracted Lawrence's most inflammatory sixteen pages for working-class readers, precisely the audience the authorities feared would be drawn towards materialism.

Never likely to become a fiery freethinking classic, the *Constitution of Man* discussed theology in reverential terms and claimed to be compatible with every word in the New Testament. Yet if a knowledge of nature's laws was not enough to ensure salvation, so too was Scripture an incomplete guide to appropriate behaviour in everyday life:

> A correct interpretation of revelation must harmonize with the dictates of the moral sentiments and intellect, holding the animal propensities in

> subordination. It may legitimately go beyond what they, unaided, could reach; but it cannot contradict them; because this would be setting the revelation of the bible in opposition to the inherent dictates of the faculties constituted by the Creator, which cannot be admitted; as the Deity is too powerful and wise to be inconsistent. But mankind will never be induced to bow to such interpretations, while each takes his individual mind as a standard of human nature in general, and conceives that his own impressions are synonimous with absolute truth. The establishment of the nature of man, therefore, on a scientific basis, and in a systematic form, must aid the cause both of morality and religion.[22]

The problem, especially for evangelical readers, was that *Constitution* maintained that understanding the laws of nature must be a preliminary to all religious instruction, so that the Bible needed to be interpreted in light of the *Constitution* rather than the other way around.

Many readers also feared that equating the mind with the brain made human spiritual destiny a predetermined question, removing any need for Christ's atonement or the doctrine of original sin. *Constitution* was denounced from pulpits, removed from libraries, and even occasionally burned. The title page of Philip Jones's *Popular Phrenology Tried by the Word of God, and Proved to be Antichrist, and Injurious to Individuals and Families* showed the book and an accompanying phrenological bust in flames (Fig. 18).[23] It did not help that Combe had derived the basic idea for his book from an unpublished tract by Spurzheim, *A Philosophical Catechism of the Natural Laws of Man*, which in turn was based on Constantin Volney's *Law of Nature; Or, Principles of Morality, Deduced from the Physical Constitution of Mankind and the Universe*. Volney's *Law* was the catechism of the French revolutionary citizen, and was frequently appended to his celebrated *Ruins*. Combe managed the presentation of religious issues with considerable skill, but his book became a target for fears that the laws of nature revealed by science would replace the need for a caring God.

THE CONSTITUTION OF MAN—HIS SOUL, MIND, AND BRAIN.

POPULAR PHRENOLOGY

TRIED BY THE WORD OF GOD,

AND

PROVED TO BE ANTICHRIST, AND INJURIOUS TO INDIVIDUALS AND FAMILIES.

BY PHILIP JONES.

"THE FIRE SHALL TRY EVERY MAN'S WORK, OF WHAT SORT IT IS."—1 *Cor.* iii, 13.

LONDON:

PUBLISHED BY SHERWOOD, GILBERT, AND PIPER,

23, PATERNOSTER ROW.

1845.

Fig. 18. The *Constitution of Man* in flames, together with a phrenological bust and other works on the science. Such images, paradoxically, may have convinced some readers that phrenology was not so much dangerous, but liberating and new. Title page of Philip Jones, *Popular Phrenology Tried by the Word of God* (London: Sherwood, 1845).

Converting the Heathen

What made *Constitution* truly terrifying in some quarters, however, was not just what it said, but the numbers that were reading it, as edition after edition poured from the Chambers's printing establishment from the mid-1830s onwards. Yet, it emerged in unpromising circumstances and with a limited circulation. *Constitution* germinated in the fervid, sectarian intellectual world of Edinburgh, in a lecture Combe delivered on human responsibility to the Edinburgh Phrenological Society in 1826.[24] In the following year about eighty copies of an expanded version were printed for private distribution within phrenological circles. This led to bitter internal infighting about Combe's religious orthodoxy, and early accusations that he was an 'infidel' for arguing that obedience to the natural laws was more conducive to human happiness than a belief in the afterlife. Combe, however, decided to go forward with publication, and *On the Constitution of Man*'s first edition was issued by John Anderson of Edinburgh as a small six-shilling book in paper-covered boards. Both Anderson and the book's printer, Patrick Neill, were keen advocates of phrenology who had been issuing Combe's books since 1823.[25] *Constitution* was distributed elsewhere in Britain by the large firm of Longman, who also had Herschel's *Preliminary Discourse* on their list.

The first edition of *Constitution* expanded on the privately circulated original both in length and ambition, but remained focused on shaping the future of the new science of mind. Combe did not compromise, and the choice presented to readers was stark. You could believe that the world, created perfect, had fallen into disorder, so that the only way for humanity to progress at all was through spiritual salvation. Or you could accept that the world as created by God was a progressive system, containing within the possibility of humans improving themselves by obeying the natural laws. Combe's insistence on this choice split the Edinburgh Phrenological Society, leading to mass resignations and angry letters in the newspapers. For Rev. William Scott and other evangelical members, Combe's attempt to draw out the consequences of a scientific knowledge of mind reeked of determinism and atheism, calling up the horrors of the French Revolution. Combe, in contrast, argued that an understanding of the natural laws was an essential prerequisite to appreciating the higher truths of Christianity. Although translated into Swedish and French, and issued several times in the United States, in Britain *Constitution* was initially a flop. Published in an edition of 1,500, the book sold slowly until 1835, about 100 copies a year, and received almost no reviews.

In 1832, however, an eccentric bequest by William Henderson, son of a wealthy Edinburgh banker, left several thousand pounds for the encouragement of phrenology. These funds were used to cut the price of the remaining 200 copies of the first edition of *Constitution* by nearly two-thirds, and an audience of medical students, shopkeepers, clerks, and mechanics who had been attending lectures by Combe in Edinburgh snapped these up as a bargain. Two revised, expanded, and subsidized 'Henderson' editions in the same format—1,000 and 3,000 copies respectively—also sold quickly.[26]

Such numbers did not go unnoticed among the printers and publishers of Edinburgh, not least because so many were also phrenologists. Among the most successful in the business was Robert Chambers, one of the brothers responsible for making the firm of W. & R. Chambers the country's leading laboratory for experimenting with economies of scale in publishing.[27] Every week the firm's celebrated steam-powered machine printed tens of thousands of copies of the penny-and-a-half weekly *Chambers's Edinburgh Journal,* which offered an attractive mix of essays, fiction, biography, and science. In the early 1830s Chambers had become a keen convert to the views on natural law expressed in the *Constitution of Man,* believing that

'Phrenology appears to bear the same relation to the doctrines of even the most recent metaphysicians, which the Copernican astronomy bears to the system of Ptolemy.'[28] As Chambers saw it, though, the key issue was not phrenology, but the naturalistic view of mind and character it involved. As he told Combe privately, the problem the phrenologists faced in converting the British people was analogous to that of the Rev. Alexander Duff in his missionary labours to the Indian subcontinent:

> With deference to your opinion, I would rather think that the metaphysics and practical efficacy of phrenology constitutes the point of the wedge which is yet to rend asunder the mass of philosophical heathenism and make way for the full reception of this great new light. Phrenology, as phrenology, has made hardly any progress since I can recollect the Edinburgh world: the circle of its adherents has not extended. But . . . present them with the philosophy apart, and they greedily receive it—yea as if it was a new gospel. . . . The process is analogous to that recommended by Duff for converting the heathen. First convince them that a correct system of mind has been discovered, and there will then be no obstacle to the reception of its fundamental truths except what may still be presented by the habit of venerating former systems. When we reflect that some of the forms of heathenism survived in Scotland till the close of the last century, perhaps eight hundred years after Christianity had been acknowledged to all intents and purposes, we must not fret at the slow progress which phrenology makes. . . .[29]

Both as businessman and advocate of the new philosophy of nature, Chambers sensed an opportunity. The sales of the 'Henderson' editions of *Constitution* suggested that an even larger audience could be reached if the price was lowered still further. This could be achieved not through charitable subsidy, but by reducing production costs to the bare minimum by exploiting the spare capacity of the firm's steam printing machine. The inspiration for this idea probably came from the United States, where earlier in 1835 William Pearson of New York City had issued a pirated version of the *Constitution of Man* as a sixty-page number in his monthly periodical, the *Alexandrian*.[30] The

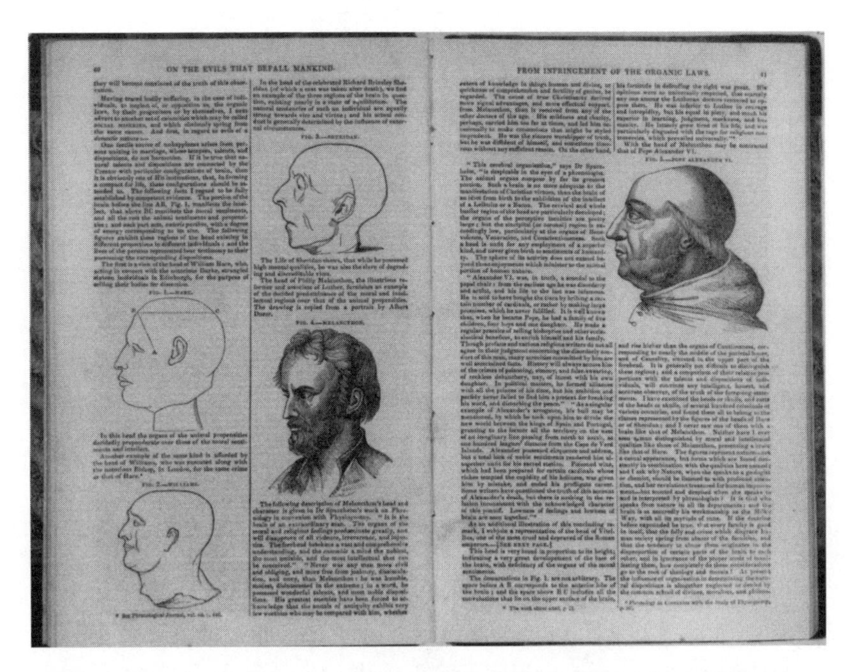
ON THE EVILS THAT BEFALL MANKIND

FROM INFRINGEMENT OF THE ORGANIC LAWS

Fig. 19. The closely printed double columns of *Constitution* from the edition printed by W. & R. Chambers made the work affordable to a wide range of readers. George Combe, *The Constitution of Man*, 4th ed. (Edinburgh: W. & R. Chambers, 1836).

Chambers's editions of *Constitution*, which crowded even more words into a similar double-columned format, were spectacularly successful (Fig. 19). Printed on the steam press and published by a consortium at the amazingly cheap price of one shilling and six pence (about a fifth of the original price), they sold 43,000 copies by the end of 1836 and 85,000 copies by 1850. By any standard, this placed *Constitution* among the best-selling books of the nineteenth century.[31]

The cheap editions of *Constitution* involved radically innovative pricing structures, well beyond anything projected in Brougham's most ambitious plans for the Society for the Diffusion of Useful Knowledge. W. & R. Chambers built on the success of *Constitution* by making it the first in a series of 'People's Editions', a range of works in history, science, biography, and other fields published at unprecedentedly low prices. In many ways, the sales of *Constitution* were the culminating point of the effects of the new technologies of print and its distribution in the first half of the nineteenth century. Other publishers, on all sides of the political and religious spectrum, followed suit. In 1845, the National Corn Law League published a four-page selection of material from Margracia Loudon's *Philanthropic Economy* to support the case for repealing the taxes on imported corn, an issue of crucial significance in the wake of famine in Ireland. This tract—by orders of magnitude the most widely distributed of her writings—featured the familiar double-columned format and relied upon economies of scale that made it possible to distribute nine million copies of this pamphlet across Britain through the new penny post.[32] Such numbers were only possible because of the experiments in cheap publishing that the Chambers firm had made a decade earlier with the *Constitution of Man.*

Constituting Readers

The opportunities afforded by machine-based printing, together with the close relations between Combe and his Edinburgh publishing friends, transformed the fortunes of *Constitution.* They also fed back into his conception of his audience and the text of the book, although this has been obscured by the sheer scale of the sales figures. The book was dramatically altered and expanded over its many editions; like many other non-fiction classics of the nineteenth century, *Constitution* was effectively not a single work, but a serial production. Because typesetting was relatively cheap, works that sold well could be continually revised, thus encouraging purchasers to purchase later editions. For example, the *Principles of Geology* went through ten editions in Lyell's lifetime, and appeared in many different forms, with some parts expanding to become the basis of separate works, such as the *Elements of Geology* and the *Antiquity of Man.* Somerville's *Connexion* also went through many editions, and although the changes were less dramatic, they kept the work broadly up to date with new findings.

Constitution sold in numbers that dwarfed any of these books, and the opponents it faced were far more vehement than those who later condemned the *Origin of Species.* In many ways, the editions from 1835 onwards were different books from the early versions. They were not only priced for, but addressed to, a wider market and readers less likely to be committed to or familiar with phrenology. The significance of the revisions is apparent from the opening of the main text. In the first edition of 1828, this had read:

> A statement of the evidence of a great intelligent First Cause is given in the 'Phrenological Journal', and in the 'System of Phrenology'. I hold this existence as capable of demonstration. By NATURE, I mean the workmanship of this great Being, such as it is revealed to our minds by our senses and faculties.[33]

In the cheaper editions from 1835 onwards, this was replaced by:

> In surveying the external world, we discover that every creature and every physical object has received a definite constitution, and been placed in certain relations to other objects. The natural evidence of a Deity and his attributes is drawn from contemplating these arrangements. Intelligence, wisdom, benevolence, and power, characterize the works of creation; and the human mind ascends by a chain of correct and rigid induction to a great First Cause, in whom these qualities must reside. But hitherto this great truth has excited a sublime yet barren admiration, rather than led to beneficial practical results.[34]

In the original passage, readers are asked to view classic theological debates through a narrow phrenological lens and are addressed as if they had ready access to the *Phrenological Journal,* which at its peak of circulation had only a few hundred subscribers. In the later and less expensive editions the narrator no longer starts off in the library, but speaks from a transcendent all-seeing perspective—familiar from Volney's *Ruins* and Davy's *Consolations*—that considers the world whole and from above. The great problem *Constitution* will address is now made evident. With passages added and reorganized, the argument broadened out, and with an entirely new chapter on science and religion, the later versions are much more effective at anticipating prejudice and introducing controversial doctrines. They illustrate the specifics of phrenology with woodcut illustrations contrasting the heads of criminals and theologians. The 'People's Edition' printed by the Chambers brothers, although occupying less than a third of the number of pages, is actually twice as long as the first edition of 1828.

These revisions invited the reader to approach the book in a new kind of way, particularly in relation to the much-debated question of whether or not phrenology was a science. In all the editions, the 'Preface' had stressed that faith in phrenological science was not a prerequisite for reading, but the new opening made this claim more convincing. Combe underlined this later in the book when he wrote that:

> In this inquiry it is indispensably necessary to found on some system of mental philosophy in order to obtain one of the elements of the comparison; but the reader, if he chooses, may regard the phrenological views as hypothetical in the mean time, and judge of them by the result. Or he may attempt to substitute in their place any better system with which he is acquainted, and try how far it will successfully conduct him.[35]

In a very practical way, the reader was invited to read at least parts of *Constitution* as though they were fictional narratives, like those that might be found in a contemporary novel, but then to realize that they are in fact grounded in truth. Even outright opponents of phrenology could benefit from this explicit blurring of generic boundaries, which had a long history in writings on controversial intellectual issues. The American abolitionist Frederick Douglass, for example, viewed phrenology as Combe's 'peculiar mental science', but praised *Constitution*.[36]

The difficulties in establishing the new science of mind were compounded by the focus on controversy about Combe's character and personality. Even among his friends, in conversation, in correspondence, and even as a lecturer, Combe had the reputation of being too much in earnest, presumably because his phrenological 'organs' of Concentrativeness, Causality, and Conscientiousness were large, that of Wit only 'moderate'. (Unlike Babbage's, Combe's organ of Number was highly deficient.)[37] As Douglass observed after a breakfast with Combe in Edinburgh, 'I was a listener. Mr Combe did the most of the talking, and did it so well that nobody felt like interposing a word, except so far as to draw him on.... His manner was remarkably quiet, and he spoke as not expecting opposition to his views. Phrenology explained everything to him, from the finite to the infinite.'[38] The actress Fanny Kemble recalled that an expert knowledge of phrenology was needed to follow dinner conversation at the Combe household, so entirely were events and actions explained on the basis of cerebral characteristics. Never a popular celebrity, Combe was often jeered at public events, so that the press took note when a crowd of three thousand applauded him for several minutes at the first soirée of the Glasgow Athenaeum in 1847, with Charles Dickens in the chair.[39]

Why, then, did so many readers find *Constitution* more compelling than its author? One way of approaching this question is to be found in Loudon's remark that the book expressed her 'inmost thoughts'. For her, as for many readers, Combe's new science meshed with a self-achieved 'common sense', with reflections she had previously expressed in her novels and in *Philanthropic Economy*. The ideas were concrete and the examples taken from everyday life: this was a philosophy that could be mastered easily, and yet gave clarity and system to views that were taken for granted across the social spectrum.

Phrenology itself, though a complex system involving nearly three dozen organs, was relatively easy to master, and related closely to assumptions about race, gender, and class: 'In a limited sense,' as one opponent, the Cambridge geologist Adam Sedgwick wrote, 'we are all of us phrenologists.'[40] Above all, *Constitution* motivated an understanding of one's own mind that was at the core of the developing idea of self-help, while showing how that understanding could be used towards broader aims of social transformation.

An essential element in the book's success was the sense the reader gained of being addressed by a calm and disinterested mentor, implicitly (although not directly) comparable with the great philosophers of antiquity. It was important that *Constitution* was not seen simply as a popularization of Spurzheim or Gall, or of other works authored by Combe, but as a book of original thought. Without making untoward claims for discovery, the novelty of separating out the three kinds of law is carefully noted. Any implications of atheism, or even a tendency to undermine faith, are quietly set aside, and clergymen are addressed as potential allies, in contrast with Combe's correspondence, where they are seen as implacably opposed to progress. The reader addressed in *Constitution* is constructed in ways that contemporaries would have recognized as the archetypal ideal of the disinterested enquirer after truth. He (for the address is consistently male) is classically educated (there are untranslated Latin quotations) and wealthy (enough to have problems in choosing servants). This may seem strange, particularly as the 'People's Editions' with their overcrowded pages must have been read primarily among the lower middle and working classes, but such readers did not wish to be spoon-fed. They did not necessarily hope to live the life implied by such modes of address, but they did want to have access to authoritative knowledge, and in the first half of the nineteenth century this had strong associations with masculine gentility, a status that guaranteed independence of thought.

In the leading middle-class weekly, *Chambers's Edinburgh Journal,* an article (anonymous, but written by the editor Robert Chambers) brought out the connection of *Constitution* with notions of appropriate behaviour. Starting out with a commonplace question, 'Is ignorance bliss?', the essay went on to suggest that it was anything but:

> The uses of studying our own nature are perhaps less obvious than those connected with the study of physical nature; but they are not less important. A knowledge of human nature, and a habitual system of acting upon it, has come to be known in this present age by the expressive word *tact*; and it is certainly an accomplishment by which many great and direct advantages are to be gained by the individual, while that welfare of others is not in the least diminished, but rather increased. There is unfortunately no *system* of knowledge of this kind, if we except phrenology, which has not yet gained general attention or credence. But every man may learn a great deal from the ever-open book of nature itself, if he is only willing to inspect the movements of his fellow-creatures with any share of attention; and, imperfect as the science is, there are many books which he may study with advantage.[41]

A footnote then led the reader to *Constitution,* remarking that the work could be read even by 'the most inveterate disbeliever in the science'. Perhaps that was the first lesson that the new science had to teach: just because you saw someone reading a book did not mean that they accepted what it said.

A Science of Tact

By the middle of the nineteenth century, aspirations to establish phrenology as a science of mind were fading. A comparison with geology, which had faced similar problems in establishing its legitimacy, is revealing. In 1830, the study of the earth still had much of the reputation for speculation and irreligion that had earlier led to its being labelled as a 'species of mental derangement'.[42] Two decades later the situation was dramatically different. The discussions of geology in Lyell's *Principles,* Davy's *Consolations,* and Herschel's *Preliminary Discourse* had contributed to securing the philosophical credentials of the new science for all but a few holdouts among Scriptural literalists and millenarian evangelicals. Combe's *Constitution* had a different fate. The specifics of phrenology suffered a range of attacks from an increasingly defined scientific community, among them the physiologist William Carpenter and the anatomist Peter Mark Roget. Forms of publication and the role of the scientific investigator of the mind were changing, as journal articles announcing new discoveries assumed a primacy that few could have predicted even a decade or two before. The *Phrenological Journal* looked increasingly out of step and ceased publication in 1847; it had helped to hold together a small group of committed enthusiasts, but its original aim to imitate the political quarterlies created a sectarian impression, and the absence of new discoveries was alarming. In 1853 the fifth edition of the *System of Phrenology* featured revisions to the sections on brain physiology, but announced the dismal news that nothing new in phrenology had been discovered for a decade—hardly a promising sign for an advancing science.[43]

In terms of sales, the *Constitution of Man* was vastly more successful, but it was increasingly read for the light it shed on the laws of nature, rather than for any specific phrenological claims about the mind. The invitation to read parts of the book as fiction was clearly being taken up. *Constitution* provided the invisible underpinning for a whole range of Victorian reform movements, from removing Church control over education to repeal of the taxes on corn. In 1854, for example, the prime minister Lord Palmerston received a petition recommending Combe's views as expressed in a recent pamphlet, which argued that 'criminal legislation and prison discipline will never attain to a scientific, consistent, practical, and efficient character until they become based on physiology'.[44] The petition had been signed by a host of medical luminaries, including some of phrenology's most vociferous opponents. Palmerston, who shared Queen Victoria's enthusiasm for the science, was favourably impressed.[45]

Constitution helped to make sense of life in an increasingly urban and individualized consumer society. It allowed readers to interpret the character of others as expressed in their behaviour, whether they were intimate relations, strangers on the street, or potential political allies. *Constitution* offered not only a way to choose appropriate servants or reform criminal offenders, but also encouraged readers to believe they could understand the springs of inner action, why people spoke and acted the way they did. The book also offered the prospect of a 'scientific' basis for distinguishing between different races, for understanding the respective roles of women and men, and for making policy to deal with the country's rapidly growing global network of trade and influence. 'The Master of the universe,' Lord Palmerston wrote, 'has established certain laws of nature for the planet in which we live, and the weal or woe of mankind depends upon the observance or neglect of these laws.'[46]

The reasons for the long-term appeal of *Constitution* are best indicated in the host of attacks on phrenological science at the time it was just coming to national prominence in the 1820s. A print made in September 1826 (Plate 6) shows a phrenological lecturer speaking to a prosperous but undiscerning audience. That the lecturer is addressing the problem of interpreting character is evident not only from the grotesquely bumpy heads of his listeners, but also from the cleverly selected shelves behind him, in which 'Gall' is in a bottle rather than a book, and the salacious *Memoirs of Harriette Wilson* sit next to *The Whole Duty of Man.* The desire to judge character through material things is brought out particularly well in the poster on the wall by the lecturer's left hand. This advertises a variety of bizarrely misshapen hats tailored to wearers with specific phrenological characteristics: 'Sold by Royal Patent: Phrenological Hats Adapted to Every Protuberance or Faculty Yet Discovered'. The theme of fashion is developed further in another print (Plate 7) entitled 'Toe-tology—versus—Phrenology.—A Course of Lectures to be delivered (at the "Foot" of Arthur's Seat.)'. Here character is shown to be revealed by the form of the foot and the shape of the shoe rather than by the head or hat. It includes 'Blue-stocking-itiveness immense', with untied lace and a hole in the stocking, and the tight-fitting 'Cheap-Shoe Shop-itiveness'. Revealingly, even political views could be inductively inferred from footwear, as in 'Anti-Reformitiveness and Club-ology prodigious'. Since voting was conducted by secret ballots, even after reform, this kind of sign could be very useful.

Such jokes worked because appearances based on clothing—or other commodities—could be so completely misleading. This was a major theme of contemporary fiction, which attracted large middle-class audiences struggling with the ambiguities of outward appearance in relation to moral, religious, and political authority. Anyone could wear sensible shoes, purchase fashionable verse, own a library of profound philosophical books, or don the bonnet of a revolutionary Jacobin. What people wore or what they read could be seen as superficial. The plot of Margracia Loudon's *First Love*, like dozens of novels in the period after Jane Austen, depended entirely upon the heroine's ability to understand what was going on behind the actions of potential male suitors. Documents were forged, incriminating documents had been written in invisible ink. How could one find out the truth in such circumstances?[47]

These themes come across strongly in *Modern Accomplishments, or, the March of Intellect*, first published in 1836 and dedicated to Princess Victoria.[48] Throughout the novel, the author, Catherine Sinclair, aimed to use fiction to show that women's education had for too long placed 'ornamental above useful accomplishments', and it is no surprise to discover that she also designed the anti-phrenological satire 'Toe-tology'.[49] Clothing is one of the ways that Sinclair reveals the character of the contrasting cousins, Eleanor Fitz-Patrick and Matilda

Howard, during the course of the narrative. Eleanor, a great beauty and slave to fashion, has feet so small that they had been 'compared to a Chinese lady's'. Brought up to think that outward show was all that mattered, at home Eleanor's hair was in papers and her shoes 'often slip-shod', but in society her dress was admired, envied, and imitated. The parties of her mother, Lady Fitz-Patrick, were the most brilliant in fashionable Edinburgh, while those of Lady Howard ('the walking Library') laboured to be the most intellectual, attended by mathematicians, phrenologists, and authors of all kinds. Their two daughters had been brought up accordingly. 'People who cannot dress themselves ought to be voted incompetent to manage their own affairs,' Eleanor tells the devout, long-suffering Matilda, 'and put into the hands of a jury of milliners.' Even their old uncle Sir Philip comes under Eleanor's critical gaze as a fashion victim, with his 'agonisingly tight shoe drawn over his gouty foot'.[50] The same point about the dangers of relying on outward appearance is underlined in the conclusion to *Modern Accomplishments*, as Matilda is happily confirmed in her Christian faith, while Eleanor makes a brilliant marriage but is constantly exposed to temptation.

In this world of masks, misleading impressions, and the clutter of material things, *Constitution* offered a way of using surfaces to penetrate to the underlying nature of individual character. The book exemplified the value of 'tact', an ability to feel one's way in a society that was increasingly anonymous, urban, and prone to individual self-interest. The Edinburgh philosopher Dugald Stewart had first characterized 'tact' in this sense in the 1790s, in referring to the way that ordinary citizens caught up in the French Revolution needed flexibility and discrimination.[51] In such settings, judging character was critical to success and even survival. To opponents, particularly those reading the book from an evangelical religious standpoint, the use of phrenology in *Constitution* could be seen as a dogmatic attempt to subvert religious faith. More commonly, though, readers believed that *Constitution* provided a scientific form of self-knowledge and knowledge of others. Combe's distinction between moral and intellectual laws, organic laws, and those pertaining to the physical world was certainly simple enough, but it offered a way for readers to think about these laws together and to apply them to their own circumstances on a daily basis. Reading, writing, and arithmetic were essential elements in education, but the modern age was thought also to require knowledge of the laws of nature. What made *Constitution* compelling was that its explanation of those laws made 'tact' scientific.

7

The Torch of Science

Thomas Carlyle's *Sartor Resartus*

> *Sartor Resartus* we like amazingly. It is as amusing a piece of German mystification as was ever penned in the King's English.
>
> —*Morning Post*, December 1833

Readers of *Fraser's Magazine for Town and Country*, the fashionable satiric Tory monthly, were startled by a serial that began in November 1833. Almost any subject they could imagine—geology, astronomy, society, even sleep and drunkenness—had been the subject of learned reflection in recent books. But *Fraser's* announced that there was still something lacking: a scientific analysis of clothes. This unpromising theme was developed in eight instalments, which purported to be based on a bizarre treatise written by a German professor of 'things in general' at the University of Weissnichtwo (know-not-where). The narrator and editorial compiler of 'Sartor Resartus: In Three Books' was anonymous, but the professor's name was Diogenes Teufelsdröckh, which combined the name of the great cynical philosopher of antiquity (usually translated as 'God-born') with the German for 'Devil's shit'.

What kind of work was this? Among many other things, 'Sartor Resartus' ('the tailor retailored') was a complete inversion of the reflective scientific treatises that flourished around 1830. Books as varied as Combe's *Constitution* and Herschel's *Preliminary Discourse* featured unifying perspectives and observable phenomena, but 'Sartor' unravelled the relation between external appearances and internal character. These works typically employed a controlled rhetorical register with subtlety and skill, but 'Sartor' was prolix, paradoxical, and personal. Those who wrote and lectured upon 'things in general' aimed to be encyclopaedic, but 'Sartor' was incomplete. The early instalments sketched editorial methods and recalled the narrator's encounters with the German professor, but these were all about difficulties and obstacles. The middle ones focused on Teufelsdröckh's autobiography, but this was unfinished. And just as the author seemed close to proposing something that looked like a profound philosophy, the final chapters turned to comic digressions on fashion-conscious dandies and lowly working tailors. Even for sympathetic modern readers, much of this has been hard to explain.

As contemporary novels and satires like 'Toe-tology' (Plate 7) suggest, readers of 'Sartor Resartus' lived in an era that was fascinated by clothes. Not only did fashion have a central role in the plots of 'silver-fork' fiction, but men and women regularly attended to the latest cut of trouser or sleeve, and the impact of machines on hand-loom weavers, tailors, and other textile workers was widely debated. Paper was made of rags, and so books and periodicals—even the pages of *Fraser's Magazine* itself—came within the purview of an analysis of clothes. Clothes were notoriously the most superficial signs of character, but for that very reason they provided a way to think about insides and outsides, appearance and reality, who someone was and how they appeared to others. However improbable, a science of clothes (or so the anonymous Editor in 'Sartor' hoped) could communicate perspectives on the world that were vital, dynamic, ironic, and profoundly serious.

Heroic attempts have been made to see 'Sartor Resartus' as a novel, as philosophy, as autobiography, or as a satiric essay in the mould of

Jonathan Swift's *Tale of a Tub* from the previous century. It draws on all these genres, but only to undermine them, to demonstrate the impossibility of drawing sharp lines between different literary forms. Like all of the books we have been exploring, 'Sartor' relates to many forms of writing, but the one to which it does so most insistently is the literature of scientific reflection in the reform era. As the American essayist and lecturer Ralph Waldo Emerson recognized, 'the manifest design of the work... is, a Criticism upon the Spirit of the Age,—we had almost said, of the hour, in which we live'.[1] It implicitly questions all the books discussed so far, and suggests that any attempt to discover meaning though a narrowly conceived physical science is bound to fail. It condemns all such efforts as variants of a barren utilitarianism, written in a mechanical style to be read in a mechanical way. In the early 1830s, this was an unusual and idiosyncratic reading of the reflective literature of science, which captures neither the diversity of its questions nor the richness of its language, yet it embodies many of our own aesthetic sensibilities and taken-for-granted divisions between 'science' and 'literature'. Against method and opposed to the 'clear' style of the reflective tradition, 'Sartor' was a call for the birth of a spiritually vital science that would release the human potential for action. As the work announced, 'there is something great in the moment when a man first strips himself of adventitious wrappages; and sees indeed that he is naked, and, as Swift has it, "a forked straddling animal with bandy legs"; yet also a Spirit, and unutterable Mystery of Mysteries'.[2]

Although the moderate Tory *Morning Post* was enthusiastic about the serial, most contemporary readers were puzzled and even angry; at least one threatened to cancel his subscription to *Fraser's*. The anonymous author of 'Sartor', Thomas Carlyle, had been a rising star both in its pages and those of the quarterly *Edinburgh Review*. (He had also written for the *Post*, which makes its unusual enthusiasm perhaps more understandable.) Carlyle, who had vacillated between publishing the work in periodical form or as a book, had fifty-eight copies of the *Fraser's* instalments bound together and sent to friends, including Emerson in Boston. Emerson arranged for publication of *Sartor Resartus* as a book in the United States, where it appeared for the first time under Carlyle's name in 1836; it was finally published in this form by Saunders & Otley in Britain two years later. After this slow start it became one of the key works of the nineteenth century, read by thousands who regarded Carlyle as the prophet of a new kind of science.

A Serial Philosophy of Clothes

Understanding 'Sartor Resartus' in the context of the early 1830s means forgetting much of what later generations of readers have made of it. As a serial in a monthly periodical, it was read in instalments, in the context of other articles in the journal, and without a knowledge of its length or ultimate shape, other than that it would be in three books.[3] The *Morning Post*, for example, had initially condemned the 'peculiarities of the style' on the basis of the first instalment and only after the second adopted its more positive view. Readers could dip in and out of an episode, consider it in relation to other articles, or ignore it. Even more significant is the fact that the work, like almost all writings in general periodicals, was anonymous. Carlyle was soon to become one of the most celebrated authors of the century, but at this time he was struggling, poor, and relatively unknown. We read 'Sartor' in relation to his private letters, *Edinburgh Review* essays, and a jokey sketch about him that had appeared in *Fraser's*, but most readers in the 1830s were not in a position to make these connections.

One of the few things that contemporary readers could be clear about was that the new work announced itself as a satire on the reflective tradition of scientific works, and of 'useful knowledge' as a guide to living. 'Sartor' is a parody, an ironic inversion of the conventions of books that claimed that science provided a path to truth and understanding. The serial opened with a section labelled 'Preliminary':

> Considering our present advanced state of culture, and how the Torch of Science has now been brandished and borne about, with more or less effect, for five thousand years and upwards; how, in these times especially, not only the Torch still burns, and perhaps more fiercely than ever, but innumerable Rush-lights and Sulphur-matches, kindled thereat, are also glancing in every direction, so that the not the smallest cranny or doghole in Nature or Art can remain unilluminated,—it might strike the reflective mind with some surprise that hitherto little or nothing of a fundamental character, whether in the way of Philosophy or History, has been written on the subject of Clothes.[4]

This was a parody of the tradition of encyclopaedic preliminary discourses and scientific treatises, clearing the ground and advertising the need for the work at hand. The image of the 'torch', derived ultimately from Plato, had been taken up in Francis Bacon's call at the opening of the seventeenth century for the empirical investigation of the natural world, and was the motto for William Hamilton's celebrated philosophical lectures at Edinburgh, which Carlyle and many other young men had attended.[5] 'Sartor' outlined how the torch had already illuminated everything from the movements of the planetary system and herring migration to the theory of rents and the philosophy of language, so that the problem of creation was now 'little more mysterious than the cooking of a Dumpling'. The life and situation of humans had been exhaustively examined in revolutionary Paris and reform-besotted London and Edinburgh, where spiritual faculties and bodily tissues had been 'probed, dissected, distilled, desiccated, and scientifically decomposed'. But clothing, 'the grand Tissue of all Tissues, the only real *Tissue*' had been ignored.[6]

This gap, 'Sartor' announces, had been filled at last not in France or Britain, but in Germany—'learned, indefatigable, deep-thinking

Germany'—by Professor Teufelsdröckh's comprehensive new treatise, *Die Kleider ihr Werden und Wirken* ('Clothes, their Origin and Influence'). In the opening instalment, the Editor explains his plans for presenting the work and the difficulties surrounding the project. His first thought had been to publish in a periodical, but this would be an ideal course only if political parties could be abolished and 'the whole Journals of the Nation could have been jumbled into one Journal, and the Philosophy of clothes poured forth in incessant torrents therefrom'. The best option had proved to be *Fraser's Magazine,* a journal known for literary experiments and verbal pyrotechnics.[7] Teufelsdröckh is recalled by the Editor as a remarkable figure, yet however profound his message, his book lacks all arrangement, without 'true logical method', so that the public, hungry for the clothes philosophy, is forced to help itself from the trough of a 'mad banquet'.[8] The second instalment in *Fraser's* attempts to make sense of this 'chaos', with its 'interminable disquisitions of a mythological, metaphorical, cabalistic-sartorial and quite antediluvian cast'. The brief chapter on 'Aprons', for example, turns out mostly to do with printing and publishing; its title refers (among other things) to the paper aprons worn by Parisian cooks, which are claimed as a new outlet for printed literature. Since recycled cloth provided the raw material for the paper in books and periodicals, aprons lead Teufelsdröckh into a satiric discussion of the British newspaper industry.[9] The contradictions, analogies, and recyclings in every sentence of these opening discussions left readers of 'Sartor' without any stable sense of reference, so that nothing was simple or what it appeared to be on the surface. The chapter on 'Aprons' anticipates the moment when the text will consume itself, in a great conflagration of the mass of printed paper blocking the channels of circulation.

The Editor who serves as the narrator in 'Sartor' is a stolid English reader, conservative in politics and anxious to get to the bottom of these mysteries. One key to understanding, Teufelsdröckh has convinced him, is to read books as an expression of their authors' experience. In contrast, works of scientific reflection, with their transcendent narrative voices, rarely featured substantial biographical information about their authors other than their names and qualifications on the title page. Instead, authors in science were known to their audience primarily as a result of contemporary celebrity culture in newspapers and periodicals, which reported on the comings and goings of great men such as Humphry Davy and John Herschel. 'Sartor' presents the separation of 'work' and 'life' that this entailed as further evidence of the fatal hand of mechanism in contemporary society. Books are written as though they are not the direct expression of the inner life of their authors, when this is what matters most.

Although the Editor recalls his own memories of Teufelsdröckh, and reports these at length, he is overjoyed at the end of the second instalment in *Fraser's* to receive substantial materials about the great man. The good news is that this is not merely a biography, but an autobiography; the bad news is that it is in six paper bags of scribbled slips. The Editor's aspiration is to tell readers everything about his subject, but the records of Teufelsdröckh's life—like those of any person—are partial and fragmentary. In the next three instalments, constituting Book II of 'Sartor', the Editor painstakingly attempts to reconstruct Teufelsdröckh's early life from these scraps. At one level, the result is the familiar story of a Romantic genius, who is deposited as a baby at the doorstep by a mysterious stranger, attends school and university, and ends up a sceptic after falling under the dangerous spell of rationalism: 'thus was the young vacant mind furnished with much talk about Progress of the Species, Dark Ages, Prejudice, and the like'.[10] Without faith or profession, Teufelsdröckh then falls deeply in love, but this proves a false journey to spiritual recovery, as 'the gay silk Mongolfier' balloon of passion explodes.[11] Bruised, bewildered, and alone, the young man undergoes agonies of the blackest doubt. 'Is there no God, then,' he wonders, 'but at best an absentee God, sitting idle, ever since the first Sabbath, at the outside of his Universe, and *see*ing it go?' Was the universe nothing but the 'boundless Jaws of a devouring Monster'?[12] Teufelsdröckh, now a nameless 'Wanderer', asserts his freedom from fear, having realized that he needs to bypass outward appearances and penetrate to the nature of things through action rather than contemplation.

The Science of Natural Supernaturalism

Even if the scraps the Editor has received are fictions, they carry the message of something symbolic and spiritual that is deeper than facts. As Carlyle would have known from widespread newspaper reports, those facts of science—and specifically the precision findings of astronomers—were becoming increasingly central to daily life in Britain during the months when 'Sartor' was serialized in *Fraser's*. In late August 1833, the Royal Astronomer, James Pond, had installed a large ball on a pole at the Royal Greenwich Observatory to communicate the time (Fig. 20). Made of metal and wood, the time ball was released down the pole at 1 p.m. every day, so that the chronometers on ocean-going ships could be calibrated, thus making it possible for them to determine their longitude at sea. As it was located on the highest turret of Flamsteed House, the ball was also visible throughout much of London, and it became a notable milestone in creating 'Greenwich time' as a metropolitan—and ultimately a worldwide—standard. When Carlyle and his wife Jane moved from the isolated farm of Craigenputtoch in Scotland to Cheyne Walk in London in the summer of 1834, their time literally changed, for they had moved from a place where time was established by local custom to a place where time was increasingly set with the precision required by commercial navigation.[13]

Fig. 20. The time ball at the Royal Observatory at Greenwich, at two minutes before one o'clock, and at one o'clock; from 'Time Ball on the Greenwich Observatory', *Arcana of Science and Art* (London: John Limbird, 1834), 13, 14.

'Sartor', published just as this innovation in the public display of precision was taking effect, depicted the significance of time in a very different way. Time and space, Teufelsdröckh acknowledged, were 'spun and woven for us from before Birth itself', and without them we would not exist. Yet time was not uniform, despite the efforts of astronomers to make every moment like the last. A rigid reliance on Newtonian assumptions about time and space, 'Sartor' insisted, could block the revelation that comes only when time stops—when past and future collapse into an eternal instant of transcendent insight, 'an everlasting NOW'.[14] Those who really knew how to tell the time would know when that moment had arrived.

In discussing Teufelsdröckh's philosophy, the Editor's resistance to the Professor ebbs away, and the chapter on 'Natural Supernaturalism' in Book II is presented as 'clear, nay radiant, and all-illuminating'. At this critical juncture, 'Sartor' is in direct dialogue with readers of reflective science, through explicit references to Somerville's presentation of Pierre-Simon's Laplace's celestial mechanics in the *Mechanism of the Heavens* and William Herschel's catalogues of stars. A reader of such books, Teufelsdröckh suggests, might speak of scientific laws that operate everywhere and at all times in 'the Machine of the Universe'. In reply, he asks what 'those same unalterable rules, forming the complete Statute-Book of Nature, may possibly be?'

> They stand written in our Works of Science, say you; in the accumulated records of man's Experience?—Was Man with his Experience present at the Creation, then, to see how it all went on? Have any deepest scientific individuals yet dived down to the foundations of the Universe, and gauged everything there? Did the Maker take them into His counsel; that they read His ground-plan of the incomprehensible All; and can say, This stands marked therein, and no more than this? Alas, not in anywise! These scientific individuals have been nowhere but where we also are; have seen some hand breadths deeper than we see into the Deep that is infinite, without bottom as without shore.
>
> Laplace's Book on the Stars, wherein he exhibits that certain Planets, with their Satellites, gyrate round our worthy Sun, at a rate and in a course, which, by greatest good fortune, he and the like of him have succeeded in detecting,—is to me as precious as to another. But is this what thou namest 'Mechanism of the Heavens', and 'System of the World'; this, wherein Sirius and the Pleiades, and all Herschel's Fifteen thousand Suns per minute, being left out, some paltry handful of Moons, and inert Balls, had been—looked at, nick-named, and marked in the Zodiacal Way-bill; so that we can now prate of their Whereabout; their How, their Why, their What, being hid from us...
>
> We speak of the Volume of Nature: and truly a Volume it is,—whose Author and Writer is God. To read it! Dost thou, does man, so much as well know the Alphabet thereof?[15]

From this perspective, modern science could never lead to genuine understanding. The answers men of science have found are like those on the 'way-bill' of a courier, charting the movements of the stars as though they were commodities; the questions that need to be asked, Teufelsdröckh demands, should echo those of God speaking to Job in the Old Testament: 'where wast thou when I laid the foundations of the earth? declare, if thou hast understanding'.[16]

How many readers of *Fraser's Magazine* would have followed this journey from the 'Woollen-Hulls of Man' to the 'Garments of his very Soul's Soul, to Time and Space themselves'.[17] Not many, was the answer from Carlyle's publisher, for the response from the public had been bewilderment and confusion. So the concluding instalment of August 1834 brought the reader back to the more familiar realm of period satire, to pose a query—'what use is in it?'—that no one who had understood the previous parts of the work would be tempted to ask. This was, again, an ironic undercutting of the scientific tradition, which 'Sartor' is depicted as being concerned only with questions of practical application. The answer is provided in two chapters devoted to the natural history of society drawn according to the esoteric science of clothes. For many modern readers, this has seemed anticlimactic, asking us to turn from the sublime philosophy of 'natural supernaturalism' to something akin to the caricatured drollery of 'Toe-tology'. Just when it looked as though 'Sartor' might have a serious, coherent message, the professor's book offers a pair of seemingly trivial chapters: one on the fashion-obsessed Dandy ('as others dress to live, he lives to dress')[18] and the other on the Tailor, considered as the unacknowledged maker of the fabric of human life.

'The Dandiacal Body' reflects on the man whose obsession with clothing epitomizes the truth of Teufelsdröckh's philosophy. As contemporary readers of 'Sartor' would have known, the Dandy was a stock figure in the early nineteenth century, part of the contemporary fascination with inner and outer meanings. 'Sartor' acknowledged the tradition of dandyism as a vital inspiration for the clothes philosophy, and used it to draw wider lessons for political economy. Acting as a scientific explorer of society, Teufelsdröckh notes that for the fashion-obsessed, all-consuming Dandy, clothes are a religion, and the sacred texts are 'Fashionable Novels' dealing with glamorous life in aristocratic society. The writers of these novels typically paraded their own status and social connections, and among the most prominent of them (and one of *Fraser's* regular targets) was Edward Bulwer-Lytton, described in 'Sartor' as 'the high priest of Dandyism'. Teufelsdröckh describes the Dandies in racial terms, comparing them with the rag-wearing, root-eating 'Poor-Slaves' of Ireland, who represent a sub-species of Drudge. The contrast between Dandies and Drudges is developed in 'Sartor' through a parody of the political economy of David Ricardo and his followers, in which the social system is described as a giant static electrical machine of the kind used in contemporary science lectures, medical treatments, and demonstrations (Fig. 21). Money moves through the mechanisms of political economy in the same way that galvanic fluids circulate in the electrical machine and into the human patient. The scientific demonstrator turns a crank that spins a glass cylinder to separate 'vitreous' or positive electricity from 'resinous' or negative electricity. Teufelsdröckh, with heavy irony, explains that the Dandy is the positive pole of the machine of political economy, attracting all the money; the negative pole is the Drudge, attracting all the hunger.[19]

The Tailor, described in a few pages of the final instalment in *Fraser's*, is clearly a Drudge in the terms of Teufelsdröckh's political economy of the electrical machine. Lowly and downtrodden, his body is so weakened by hunger that proverbial wisdom had it that 'nine tailors make a man' (Fig. 22). He could be ignored by his employers, as in 'The Cut Celestial' (Fig. 10), which caricatures a fashionable dandy with an overdue bill refusing to recognize his tailor on the street because of the difference in their social status. The Tailor lives on 'cabbage', the leftover bits of cloth which he purloins as a prerogative of the trade. Condemned by contemporary society to live as only the 'fractional part of a Man', the Tailor is without a body of substance, being made up only of the ragged clothes characteristic of his trade.

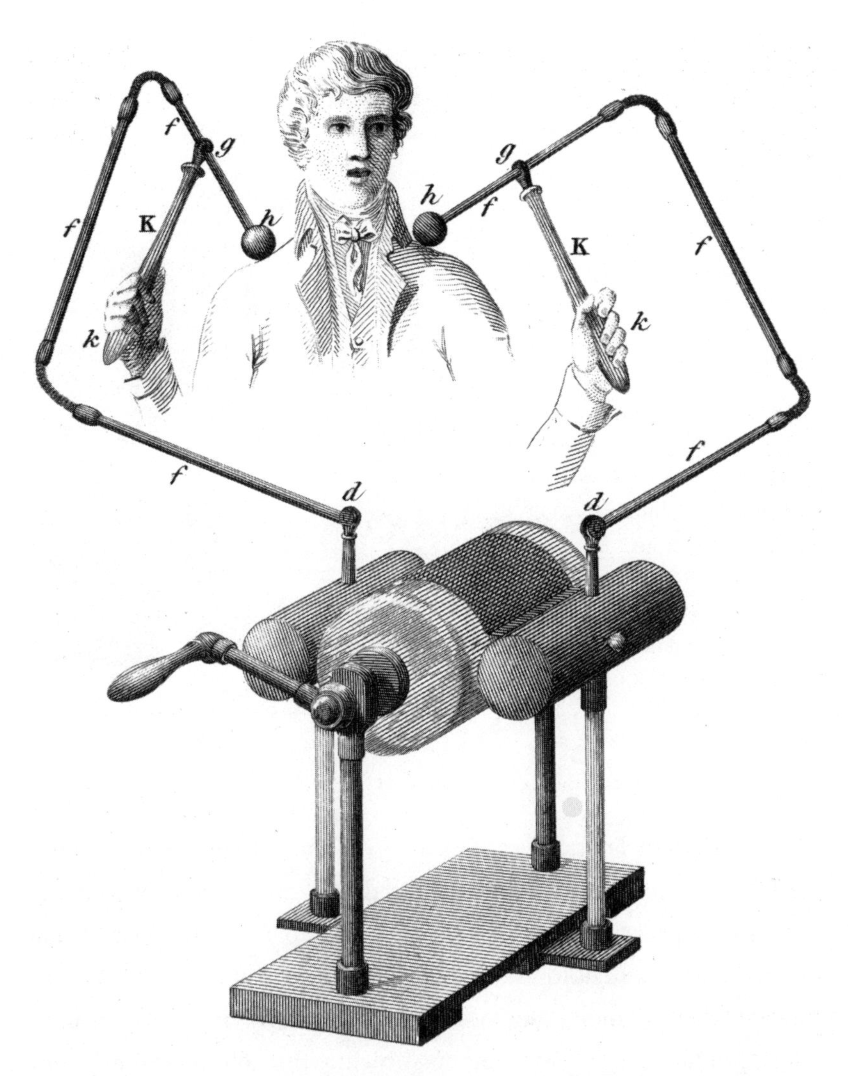

Fig. 21. Static electrical machines were commonly used at the time of 'Sartor Resartus' for public lectures and medical purposes. This form of the machine, invented in the late eighteenth century by the scientific instrument maker Edward Nairne, was capable of producing both positive and negative electricity. The operator turns the crank, creating friction between the glass cylinder in the centre and the two outer cylinders. In this illustration from 1818, the charges thus generated are being applied by a patient who holds the conductors with wooden insulating handles. 'Electricity', *The Cyclopaedia; Or, Universal Dictionary of Arts, Sciences, and Literature*, 45 vols (London: Longman, 1820), plates vol. 2, X, fig. 13.

Fig. 22. In a bad pun on 'Hungerford', a starving tailor, a young boy, and a dog look longingly at a butcher's stall. Tailors were noted for their poor diet (mostly cucumbers in the summer) and their feeble bodies. 'Hungerford Market' from Thomas Hood, *Comic Annual* (London: Charles Tilt, 1834), facing p. 132.

Most readers of *Fraser's*, and almost certainly Carlyle himself, would have been familiar with the long tradition of prints in which the outward appearance of dress signified occupation: druggists made of medicines, conchologists made of shells, entomologists made of bugs, learned women made of books. Those involving tailors had an overt political dimension, either showing threatening 'Flints' who organized themselves into clubs to resist their masters, or else as tamer 'Dungs' who did not (Plate 8).[20]

The discussion in 'Sartor' of the political economy of the tailoring trade could scarcely have been more timely. In April 1834 the 'Flints'—perhaps as many as 20,000—had gone on strike, aiming to combat the decline in conditions and attacks on their status as artisans. By early June, however, the strike had collapsed, a victim of the failure of efforts to create a general union among all workers. 'Sartor' goes even further than the striking tailors had done in asserting the nobility of labour, but does so from the starting point of radical individualism. In Teufelsdröckh's critique, the first step in achieving power will not be through socialism or joining a union, but through self-realization in useful work, the highest vocation to which humanity can aspire. In the philosophy of clothes, the tailor is a god, standing as the symbolic manifestation of the creator. An active maker, he is a whole man, touched with the divine. This is explained in 'Sartor' by a mathematical analogy, which gives a one-dimensional man we pass on the street the breadth and depth of spirit: 'the Fraction will become not an Integer only, but a Square and a Cube'.[21] As Teufelsdröckh had put it in an earlier chapter:

> 'To the eye of vulgar Logic,' says he, 'what is man? An omnivorous Biped that wears Breeches. To the eye of Pure Reason what is he? A Soul, a Spirit, and divine Apparition. Round his mysterious Me, there lies, under all those wool-rags, a Garment of Flesh (or of Senses), contextured in the Loom of Heaven; whereby he is revealed to his like, and dwells with them in Union and Division; and sees and fashions for himself a universe, with azure Starry Spaces, and long Thousands of Years. Deep-hidden is he under that strange Garment; amid Sounds and Colors and Forms, as it were, swathed in, and inextricably over-shrouded: yet it is sky-woven, and worthy of a God. Stands he not thereby in the centre of Immensities, in the conflux of Eternities?'[22]

The man who works with cloth epitomizes the power of labour in the world.

A Mathematical Apprenticeship

The divine status of the tailor is most apparent in what 'Sartor Resartus' presents as the inspiring spark of the clothes philosophy. Walking in Edinburgh, Teufelsdröckh claims to have seen a tailor's signboard showing a pair of breeches, between the knees of which were the words 'Sic itur ad astra' ('thus one rises to the stars'), an inscription from Virgil's classical epic, the *Aeneid.* This ironic moment of romantic inspiration not only embodies the philosophical essence of 'Sartor' (making breeches is the way to an understanding of the universe) but also something of Carlyle's own biography.[23]

Carlyle's first job was teaching mathematics at the Annan Academy, a school for boys in Scotland, and from there he went to study at the University of Edinburgh, where he read divinity and excelled in mathematics. 'For several years,' he recalled at the end of his life, 'geometry shone before me as the noblest of all sciences, and I prosecuted it in all my best hours and moods.'[24] When the Carlyle's name appeared in print for the first time in 1817, it was as the 'ingenious young mathematician' who had given a solution to a geometrical problem posed in a standard textbook by John Leslie, Playfair's successor as professor of mathematics at the University of Edinburgh. Carlyle was thus trained in the same mathematical traditions that had guided Mary Somerville, both in her early studies and in her redaction of Laplace's *Mécanique céleste.* In the common sense philosophy Carlyle learned at Edinburgh, mathematics was prized as the standard of truth. Leslie, for example, rejected the validity of negative and imaginary numbers, believing that geometry provided the surest basis for certain knowledge and moral standards. Carlyle was one of Leslie's best students, and after finishing his studies took up a position as a young schoolteacher in Dundee, where he engaged in mathematical duels in the local newspaper. He also had a keen interest in the sciences, and learned German to read geology. Putting his skill with languages to use, he undertook a translation of an article on chemical affinities by the celebrated Swedish mineralogist and chemist Jacob Berzelius. This paid literary work was the first of a series of projects he undertook for the 'very shifty' overworked journalist, natural philosopher, and writer on scientific decline David Brewster, including translations for the *Edinburgh Philosophical*

Journal and at least twenty articles in the *Edinburgh Encyclopædia* on topics ranging from the French essayist Montesquieu to the history of Persia. Carlyle's first book-length translation, also commissioned by Brewster, was of Adrien-Marie Legendre's *Elements of Geometry and Trigonometry*, which appeared in 1824.[25]

Carlyle's ability to mock the developing traditions of scientific writing thus drew on long experience of teaching, translating, and reviewing. His early commissions, undertaken at speed and often with a bad stomach, required mastering the conventions of prose that readers would find clear, concise, and accessible. The paid writer's job was to be invisible, so that even the translation of Legendre's textbook, a complex mathematical work with fresh material, appeared without Carlyle's name. Carlyle knew that genuine skill was required in such tasks, recalling this work as 'right enough and felicitous in its kind!' To the end of his life he remained proud of the introductory discussion on 'Proportion' that he had added to Legendre's *Elements*, knowing this topic was not usually included in treatises of this sort. Contemporary students at the University of Cambridge—one of the few places where this material was taught—were curious to know who might have written it.[26]

As a young man, Carlyle had anticipated that science would lead towards 'sublime mysteries', so that mastering Laplace's views of the astronomical system would open up the foundations of human understanding. Even at this stage, however, recognition of the practical utility that mathematics might have in making a career as a teacher and tutor was tempered by growing doubt about its significance as a route to self-realization. He had explained to a friend in 1818 his disappointment in mathematical study:

> Surely it *is* a powerful instrument which enables the mind of a man to grasp the universe & to elicit from it & demonstrate such laws—as that, whatever be the actions of the planets on each other, the mean distances of each from the sun & its mean motion can *never* change: and that, *every*

> variation in *any* of their elements must be periodical. To see these truths... to *feel* them as one does the proportion of the sphere & cylinder! 'Tis a consummation devoutly to be wished—but not very likely ever to arrive. Sometimes, indeed, on a fine evening... I say to myself—away with despondency—has thou not a soul and a kind of understanding in it? And what more has any analyst of them all?[27]

Carlyle's disillusion comes across most powerfully in an article for the *Edinburgh Encyclopaedia* on Blaise Pascal, the seventeenth-century French philosopher, mathematician, and moralist. The article praises Pascal for combining faith in religion with profound mathematical intuition and scientific experiments on air. But it also notes that for Pascal faith stood outside the 'empire of reason'. By the time he wrote this article in 1822, Carlyle had come to believe that a spiritually valid moral and philosophical understanding would not come from science as currently practised. However useful and necessary, mathematics was relegated to the role of a calculating machine that grinds out its results mechanically. The model for this was Pascal's 'famous arithmetical machine', a device that could add and subtract two numbers directly, as well as being able to multiply and divide by repetition (Fig. 23). Carlyle's encyclopaedia article derided this as 'a wonderful but useless proof of its author's ingenuity'. The image of the calculating machine appears repeatedly in Carlyle's writing, often tied to the industrial image of a grinding 'mill', which simultaneously played on the name of the well-known political economist James Mill, father of the philosopher John Stuart Mill. As Carlyle put it in the *Edinburgh Review*, 'Without undervaluing the wonderful results which a Lagrange or a Laplace educes by means of it, we may remark, that their calculus, differential and integral, is little else than a more cunningly-constructed arithmetical mill.'[28] There had been greatness in mathematical discoverers, as illustrated by Pascal's early mathematical intuitions, but not in the present age. The 'arithmetical

Fig. 23. 'Machine arithémetique', Blaise Pascal, *Oeuvres*, ed. Charles Bossut, 5 vols (The Hague: Detune, 1779), vol. 4, plate 1.

machine' had won, so that turning the handle was now all that was required.

As part of his search for a vocation, Carlyle struggled to write a novel in the 1820s, but these efforts failed, and he began to believe that his strengths would be in history, biography, and social criticism. A growing passion for German literature, first sparked when learning the language to pursue mineralogy, led Carlyle to pursue a hand-to-mouth existence in order to introduce the culture of Goethe, Schelling, and other writers to the British Isles. Their ideas provided a stark contrast with what Carlyle had come to see as the desolating influence of utilitarian philosophy in Britain, where even the mind had become 'a sort of thought-mill to grind sensations into ideas'. 'Grind' is a word that appears frequently in this context. Carlyle diagnosed the problem by distinguishing between 'mechanics' and 'dynamics' in the study of morals, a comparison that was itself drawn from the physical sciences. Mechanics was limited, focused on applications, and based on experiment and observation; dynamics was primary, vital, and grounded in intuition. Carlyle's most powerful essays in the *Edinburgh Review*, particularly 'Signs of the Times', attempted to draw out the consequences of a world dominated by machines and mechanical thinking.[29] Pascal's entertaining but trivial device, he feared, was proving a harbinger of the shape of things to come.

For Carlyle, the reformers of science and mathematics appeared to be turning so-called 'useful knowledge' into a Pascal-like engine for the mechanical transformation of every area of life. In his view, a new philosophy and a new science was needed, one that elevated the divine spark in human existence and did not destroy wonder. 'Shall your Science,' Teufelsdröckh exclaims in 'Sartor',

> proceed in the small chink-lighted, or even oil-lighted, underground workshop of Logic alone; and man's mind become an Arithmetical Mill, whereof Memory is the Hopper, and mere Tables of Sines and Tangents, Codification, and Treatises of what you call Political Economy, are the Meal? And what is that Science, which the scientific head alone, were it screwed off, and (like the Doctor's in the Arabian Tale) set in a basin to keep it alive, could prosecute without shadow of a heart,—but one other of the mechanical and menial handicrafts, for which the Scientific Head (having a Soul in it) is too noble an organ? I mean that Thought without Reverence is barren, perhaps poisonous; at best, dies like cookery with the day that called it forth; does not live, like sowing, in successive tilths and wider-spreading harvests, bringing food and plenteous increase to all Time.[30]

The great natural philosophers of the past, such as Johannes Kepler and Newton, did not grind away at particular results like their modern successors, but had proceeded through 'silent meditation'.[31] For them symbols were not just to be manipulated according to rules, but to be contemplated for their inner depth and meaning. 'Sartor' is thus a plea not to get rid of science, but to turn it back to its true purpose.

Carlyle's fear was that science was being equated with mechanism. As he had written in 'Signs of the Times', 'No Newton, by silent meditation, now discovers the system of the world from the falling of an apple; but some quite other than Newton stands in his Museum, his Scientific Institution, and behind whole batteries of retorts, digesters, and galvanic piles imperatively "interrogates Nature",—who, however, shows no haste to answer.'[32] This sweeping criticism referred not so subtly to figures such as Davy, famous for his demonstrations of the battery in the lecture theatre of the Royal Institution in London, but also to his successor, Michael Faraday, just then opening up new discoveries in chemistry and electricity. Carlyle condemned Laplace as a sort of chief grinder, and in private letters later described Mary Somerville as 'an unblameable, unpraisable *canny* Scotch lady, of intellect enough to study Euclid, and not more than enough'.[33] On the basis of the *Principles of Geology*, he dismissed Charles Lyell as 'a twaddling circumfused ill-writing man' with 'bead-eyes' and an 'uninspired voice'.[34] The height of absurdity, as Carlyle told his brother in 1831, was reached when a clutch of scientific men accepted knighthoods. In Carlyle's view Herschel, Leslie, and Brewster were nothing more than charlatans playing with mechanical toys, and for them to gain honours (on the recommendation of Brougham, no less) was a sure sign of the ruling humbug: 'Quack, quack, quack!'[35]

The one man of science who really fascinated Carlyle was Pascal's successor in building mechanical calculators, Charles Babbage, whose election plans and writings—particularly *On the Economy of Machinery and Manufactures*—he followed with interest. But the fascination was that of a serpent. A letter of 1839 described Babbage as a 'mixture of craven terror and venomous-looking vehemence; with no chin too: "cross between a frog and a viper", as somebody called him; forever loud on the "wrongs of literary men", tho' *he* has his £20,000 snug! I advise men to keep off that man.' A year later Carlyle commented again on Babbage's 'frog mouth and viper eyes, with his hidebound wooden irony, and the acridest egoism looking thro' it'.[36] Babbage was the antithesis of everything Carlyle was striving for, the only public figure who actually did want to make intelligent machines. The only thing the two men shared, late in life, was a hatred of street noise and organ grinders. In his dislike of Babbage, Carlyle would have felt at home in *Fraser's*, whose readers had been treated to a devastating attack on the *Economy of Machinery* by William Maginn, the hot-tempered Irish-born editor, just three months before 'Sartor' commenced serialization. This review considered Babbage's book appropriate only for 'a lady's boudoir' or for 'a Manchester manufacturer' with 'a tumbler of gin twist'. Maginn also condemned the work as politically obtuse, both about reform generally and the decline of science specifically. Old ideas, institutions, and traditions were being rejected without recognizing their value. 'An opinion is not like a cast-off coat,' Maginn noted, 'the worse being old, but the very reverse, provided it be a right one.'[37]

For Carlyle, machines were at their most pernicious in infiltrating the sciences of human life. In the Scottish philosophical tradition in which he had been trained, the 'sciences of man' were modelled on mathematics and astronomy, as the most mature and secure of all the forms of human knowledge. If these subjects lost their way, however, this meant that the ability to understand human affairs would also be undermined. Carlyle saw signs everywhere that this was occurring. He hoped George Combe would win the contest for the Edinburgh chair of moral philosophy, for the absurdity of appointing a phrenologist to such a post would strip bare the absurdity of the mechanical basis of the whole philosophical enterprise as pursued in the Scottish capital. At least Carlyle felt he could condemn Combe's views as straightforwardly materialistic. Late in 1831 the editors of the *Edinburgh*

Review asked Carlyle to discuss the author and designer Thomas Hope's controversial *Essay on the Origin and Prospects of Man* in a broader essay that the former was preparing. Hope's complex system of transcendental metaphysics equated matter and spirit, and on that basis argued for the transmutation of species and new views of political economy and poor relief. After a long search for a copy of Hope's elusive work, Carlyle denounced it to the *Edinburgh*'s editor as the 'highest culminating-point of the Mechanical Spirit of this age; as it were, the *reductio ad absurdum* of that whole most melancholy doctrine'.[38]

Fraser's and the Attack on Useful Knowledge

At the time Carlyle wrote those words, his pessimism about the state of the nation had been reinforced by the dispiriting experience of trying to find a publisher for 'Sartor Resartus', which he had just completed. The world of bookselling and publishing was as completely corrupted by mechanism as the rest of British society. 'Literature, too,' he had earlier noted, 'has its Paternoster-row mechanism, its Trade dinners, its Editorial conclaves, and huge subterranean, puffing bellows; so that books are not only printed, but, in a great measure, written and sold, by machinery.'[39] Of course, Carlyle had always known that the project of writing a mock philosophical treatise on clothes was risky. As he announced to himself in his journal for September 1830, 'I am going to write—Nonsense. It is on "Clothes".' In November he sent the manuscript to *Fraser's* as a two-part article, but after second thoughts withdrew it for expansion and separate publication as a book. That manuscript, however, was rejected too, first by Murray and then by Longman and other publishers. Only in 1833 did Carlyle approach *Fraser's* again with his 'Satirical Extravaganza on Things in General'.

Among the plethora of periodicals in the 1830s, *Fraser's* was the most likely to accept an oddity like 'Sartor'. Founded in 1830 as a breakaway from *Blackwood's,* it too was a monthly focused on politics, religion, and social criticism. The monthlies had risen rapidly in significance around this time, largely because of their ability to respond to changing political circumstances. As the weaver-turned-journalist Robert Mudie had written in his description of London, *Babylon the Great,* 'monthly magazines have always been a sort of dripping-pans to catch whatever the fire of genius might separate from the greater messes during the process of intellectual cookery'.[40] Many features of 'Sartor' would have been familiar to readers of such magazines, which shared a reputation for extravagance, experiment, fantastical dialogues, and bantering sarcasm. *Fraser's* was not primarily a journal of literature with poems and serialized novels, but a forum for experimenting with literary devices to probe the state of the nation. The use of an editorial persona was central to *Fraser's* and *Blackwood's,* and the humour was cruel and crude. Few doubted the brilliance of *Fraser's,* but many, as the liberal *Liverpool Mercury* put it, regretted that its talent was 'too frequently perverted to very bad purposes', as when it attacked the political economist and author Harriet Martineau for the fault of being ugly.[41]

Politically, *Fraser's* favoured 'established institutions', though not the bayonet-edged Toryism of the war hero and prime minister, the Duke of Wellington. Every issue had articles mocking radicals and the reforming Whigs, especially their support for utilitarian political economy, which was seen to undermine traditional impulses of common charity. The fundamentals of the campaign had been laid out in a critical article 'On the March of Intellect, and Universal Education'.

> What are we to do in these piping times of peace? We are to follow up the mania for educating the people. General and scientific *education* is now the thing; only give the common people knowledge—this is the prevailing palladium. In fact, say its advocates, we cannot help their being learned if

> we would. The schoolmaster is abroad, and no body sent him. It was the special hand of Heaven, no doubt; no, this is not the way these gentlemen consider human affairs—it is the progress of *reason*, legitimate reason, and truth.[42]

Like so many contemporary critiques and caricatures, the essays in *Fraser's* railed against the utopian impracticality of schemes for universal education, which reviewers saw as the latest in a long line of impractical projects and financial scams, epitomized by the notorious South Sea Bubble in the early eighteenth century, and efforts to promote 'the perfectibility of man by means of "pure reason"'. On this account, even opponents of the education mania were caught in its web; fearing that Brougham and his supporters had failed to attend to moral and religious concerns, they simply projected still more educational publications and institutions. As a result, in less than three years two universities had been founded in London: the 'godless' London University and its orthodox Church of England antidote, King's College on the Strand.

Fraser's also claimed that the onward march of intellect encouraged working people towards impractical aspirations, thereby endangering traditional values which had held communities together for centuries. Every father was becoming an educational projector for a favourite son. Instead of following the 'honest callings' of the artisanal trades, young men were encourage to study 'philosophy, and philology, and pneumatology, and geology, and mineralogy, and meteorology, and pathology, and osteology, and perhaps craniology, and all the ologies and onomies, and ographies, and aulics, and draulics throughout the immense field of science'.[43] The result was at best a waste of time and at worst a recipe for discontent and envy. Moreover, the schoolmaster peddling reason could be succeeded by the demagogue preaching irreligion and democracy. *Fraser's* warned that just as socialists argued for an equal distribution of wealth and land, so too were Brougham and the Society for the Diffusion of Useful Knowledge aiming to achieve a levelling of 'the riches of the intellect'.[44]

Articles such as these, condemning the mechanization of contemporary society, made the pages of *Fraser's* a suitable setting for 'Sartor'. As Carlyle noted in offering the manuscript to the publisher, the benevolent but bewildered Editor of the clothes philosophy could serve as a good stand-in for a typical *Fraser's* reader, who he pictured as a crusty Tory antiquarian ill at ease in the modern world. Such readers, he hoped, would find that their assumptions and misunderstandings of Teufelsdröckh's often radical views were anticipated by the Editor, 'the main actor in the business'. As a serial in *Fraser's*, 'Sartor' was surrounded by articles that gave precision to its target and sharpened the parody. The absurdity of an encyclopaedic clothes philosophy masquerading as 'useful knowledge', illuminated by the 'Torch of Science' was all the more obvious in a magazine that included essays on 'The State and Prospects of Toryism' and critiques of Bulwer-Lytton's fashionable novels. The satirical aims of 'Sartor' were less topical and more fundamental than those in Maginn's fiery editorial diatribes, and its rhetoric more unconventional and difficult to follow, but the references to mechanics' institutes, steam-powered ballot-boxes, and Murray's Family Library—to mention just a few of their shared preoccupations—were obvious enough. Overall, the self-conscious rough-and-tumble of the politics of knowledge in *Fraser's* suited 'Sartor'.

In terms of form, there were additional advantages. Carlyle acknowledged in submitting the manuscript that 'Sartor' was probably more digestible in monthly doses. As a separately published book, 'Sartor' would have had to establish its satirical conventions all on its own, which in the tough publishing market of the early 1830s made success even more unlikely than it proved to be. And although the instalments were ignored by many readers, they did go out to some 8,700 subscribers—if the editor's circulation estimates are to be believed—far more than would have been reached by separate publication.

Even as a serial, however, the journal's publisher reported that 'Sartor' met with 'unqualified disapproval'. The philosophical views in 'Sartor' were too idiosyncratic to intervene in contemporary political debate, and the style too complex for the magpie readers of a monthly, even one so lively as *Fraser's*. On April Fool's Day in 1834, the *Sun*, a reforming paper known for incisive literary commentary, praised 'the thought and striking poetic vigour' of passages in 'Sartor', but thought the whole work 'a heap of clotted nonsense' and some of the writing 'a bray worthy of the most sonorous animal that ever chewed the thistle'. The problem was an overwrought and wilfully unintelligible style, which reminded the reviewer of a character in Thomas Love Peacock's comic novel *Nightmare Abbey*. The 'blooming and accomplished' young heroine, Marionetta Celestina O'Carroll, had admitted her inability to appreciate 'metaphysical subtleties' to the philosopher Mr Ferdinando Flosky, whose appetite for obscure Teutonic verbiage was based on Samuel Taylor Coleridge. Mr Flosky replies:

> Subtleties! my dear Miss O'Carroll, I am sorry to find you participating in the vulgar error of the *reading public*, to whom an unusual collocation of words, involving a juxtaposition of antiperistatical ideas, immediately suggests the notion of hyperoxysophistical paradoxology.[45]

Tongue in cheek, the *Sun* quoted part of Flosky's response as the work of 'an eminent transcendental philosopher of the day', knowing that its readers would recognize the allusion to the novel. In the newspaper's view, the author of 'Sartor' was no more mindful of the needs of ordinary people than the absurdly prolix Mr Flosky, a character in a satiric fiction. 'These two scribes would be harnessed together in a donkey cart,' the paper commented. 'They have both an equal length of ear.'[46]

The difficulty with 'Sartor'—as the controversial historian Henry Milman had lamented in reporting to Murray on the manuscript—was that the work lacked 'tact'. Its sense of humour was heavy and the prose laden with Germanic expressions and abrupt changes in tone. Readers of a monthly looked for a lightness of touch, a way of drawing conclusions subtly and without fuss. Murray reiterated the point when passing on Milman's comments to Carlyle, remarking that 'only a little more tact' was needed to produce a book at that would sell. 'Sartor' lacked the accessible, easy manner that was widely recognized as an ideal for expository prose, and particularly for writings on science. Carlyle had mastered these conventions in his early reviews and in the essay on 'Proportion', which gave readers a sense that they could accommodate difficulties and make rational sense of the world around them. 'Sartor', in contrast, smashed these conventions, in an action that was nothing less than the literary equivalent of machine-breaking. In 'Sartor', Teufelsdröckh is made to recall his schooling in the mechanics of literary composition:

> 'My Teachers,' says he, 'were hide-bound Pedants, without knowledge of man's nature, or of boy's; or of aught save their lexicons and quarterly account-books. Innumerable dead Vocables (no dead Language, for they themselves knew no Language) they crammed into us, and called it fostering the growth of mind. How can an inanimate, mechanical Gerund-grinder, the like of whom will, in a subsequent century, be manufactured, at Nürnberg, out of wood and leather, foster the growth of any thing; much more of Mind, which grows, not like a vegetable (by having its roots littered with etymological compost), but like a Spirit, by mysterious contact of Spirit; Thought kindling itself at the fire of living Thought? How shall *he* give kindling, in whose own inward man there is no live coal, but all is burnt out to a dead grammatical cinder?'[47]

To show what has been lost through strict application of the rules of grammar, the passage pushes language to its limits by moving through a dense thicket of metaphor and allusion. The teachers are 'hide-bound', which makes them personifications of the dull, leather-covered books which are all that they know. Their 'Gerund-grinder' which, like the famous medieval clock of Nürnberg, the 'Männleinlaufen', works only according to mechanism, is contrasted with the organic growth in a garden ('littered with etymological compost'); but even a garden can only produce unthinking vegetable life. What is needed is spirit, 'Thought kindling itself at the fire of fire of living Thought'. As appreciative critics such as Ralph Waldo Emerson and John Stuart Mill recognized, this kind of writing did not make many compromises with the readers of a bold but ultimately superficial monthly magazine.

The Scientific Watch-Tower

It was only after being published as a book that *Sartor Resartus* became a significant marker on the Victorian landscape. This not only involved a change in the form of publication, but also transformed how the work would be read. Instead of appearing as an obfuscating serialized commentary on the 'spirit of the age', it was approached as a spiritual biography. This new reading gave priority to the central chapters on the life of Teufelsdröckh, culminating in the exposition of 'natural supernaturalism'. Reading the work as the testament of an individual soul was facilitated by removing the text from the satiric, ephemeral setting of *Fraser's* and rendering it as an independent work that required close reading and study—a kind of literary discipleship, often undertaken after mastery of other, more approachable works. From this perspective, knowing the work was by Carlyle made a crucial difference; these were the guiding words of a sage, even if a bracingly ironic and playful one. The geologist Joseph Beete Jukes thought Carlyle had driven 'a good broad harpoon deep into the sweltering sides of the floating carcass of Humbug'.[48] From the 1840s onwards, *Sartor* became a spiritual guide for thousands of readers in Europe and America, especially young men in search of a creed to replace traditional Christianity. 'I fancied myself a Teufelsdrockh', one late Victorian cabinet-maker recalled, who read *Sartor* along with Samuel Smiles's *Self-Help*.[49]

The naturalist Charles Darwin later said that he had never met anyone less suited to scientific enquiry than Carlyle, and in an obvious sense this is true. But for many of Darwin's associates, particularly those who came to maturity in the 1840s, reading *Sartor* was a defining moment. This new generation appreciated Carlyle's parody of the visions of science from the early reform era, which looked increasingly old-fashioned. They saw his preference for a dynamic science over a mechanical one as a prophetic insight into the direction of science itself. The new stress on the biographical chapters became crucial here, as the doubts, tragedies, and ultimate resolution in the life of the German professor at Weissnichtwo became a way of working through private crises of faith and uncertain scientific careers. The ironies and contradictions of *Sartor* appealed especially to those who hoped to find a vocation in the study of nature in a rapidly changing economy. The surgeon and aspiring naturalist Thomas Henry Huxley had read many of Carlyle's works, but *Sartor* became his constant companion, developing his hatred of posturing and his sympathy for the poor. As he later recalled, 'passing from the current platitudes to Carlyle's vigorous pages was like being transported from the stucco, pavement, and fog of a London Street to one of his own breezy moors'.[50] The Irish-born experimenter and lecturer John Tyndall viewed Carlyle's writings as 'a beacon to guide me amid life's entanglements'. As a young man he discovered in *Sartor* a way to reconcile his quest for spiritual fulfilment with a belief that science held the key to the future, in which the study of nature was the study of the vestments of the divine. Through his reading, Tyndall came to reject the doctrines of dogmatic evolutionary materialism as well as traditional natural theology, with its stress on the 'utility' of a well-engineered creation. As he wrote to a friend, 'with Carlyle the universe is the blood and bones of Jehovah—he climbs in the sap of trees and falls in cataracts'.[51] Through *Sartor*, readers like Tyndall saw that scientific enquiry could be a high calling, one that could continue to merit the term 'natural philosopher' rather than 'scientist'. Science need not be reduced to a bureaucratic obsession with record-keeping and the marking of time for the commercial economy—a vision of Victorian science that was embodied in the precision astronomy of the Royal Greenwich Observatory. The evolution of matter and of life need not lead to a world devoid of spirit and governed solely by material processes.

For such readers, *Sartor* redefined the place from which true knowledge begins as the 'scientific watch-tower'. This is where 'abstract thought can take shelter', the private residence and observatory of Teufelsdröckh in the attic of the tallest house on 'what might truly be called the pinnacle of Weissnichtwo'. Busy social settings are important in *Sartor*—the street, coffeehouse, museum—but real understanding begins with the individual in isolated contemplation, distanced from society but still able to observe it. From this starting point, the natural philosopher of the future is invited to become the night watchman of society, able to speculate 'fearlessly towards all the thirty-two points of the compass', and 'tell the Universe, which so often forgets the fact, what o'clock it really is'.[52] The watch-tower is the only place in the book described in any detail, so that the reader learns that its sitting room has three windows, the bedroom offers two, the kitchen two more, together giving a perspective on 'the whole life-circulation of that considerable City'.[53] The reader is invited into this book-filled space, which is at once an intimate domestic interior and a scientific instrument. In describing it early in the work, Carlyle could have used the word 'specula', the Latin word for watch-tower, but the term employed is 'speculum', the term used for the mirror in a reflecting telescope. There is no reason to think that this was a mistake on Carlyle's part, for from the watch-tower the natural philosopher can see both near and far, and be capable of understanding 'the raging, struggling multitude here and elsewhere'.[54] The new place of science in *Sartor* is a spiritual observatory.

Epilogue

> Great must be the rewards of learning, for many are the hands which are feeding and stuffing that great blubbery brat, the Public.
>
> —*Athenaeum* (21 July 1830)
>
> Where there is no vision, the people perish. —Proverbs 29: 18

It is hard to recapture the intense enthusiasm for the new literature of science in the early industrial age. There can be no doubt that the books discussed here fired the imagination of a generation that believed that science was on the verge of transforming the human condition. Seeing the first number of the Society for the Diffusion of Useful Knowledge's *Penny Magazine* in the window of a shop in 1832, the farmboy James Croll bought a copy and was launched on a lifetime of study. The opening article, on 'Reading for All', explained how the weekly issue was the equivalent of the finder on a telescope, because it would make it possible to locate afterwards 'more elaborate and precise knowledge' in books. Croll used the magazine in just that way, reading widely in astronomy and natural philosophy and ultimately becoming a leading authority on cosmology, even while earning his living as a janitor. The mechanic Timothy Claxton wrote that those who had learned the benefits of scientific knowledge had a duty to go forth as its missionaries: 'there is nothing that we could do that would have so great a tendency to alleviate the miseries which are felt by the great bulk of mankind'.[1]

Not only in Britain but throughout much of the Continent and in America, the late 1820s and early 1830s were a moment when dramatic change seemed possible and even probable. The movement towards political and religious reform is only the most evident of a series of efforts that have come to define a pivotal point in history. A machine-based revolution in communication, although it took decades to come to fruition, was, for the first time, clearly on the horizon. In the preface to *Lyrical Ballads* in 1802, the poet William Wordsworth had contemplated as a distant possibility the potential of science for shaping everyday life.[2] A quarter of a century later that age of science appeared imminent. The decade around 1830 was a period of projections, projects, and prophesies, of attempts to imagine the future. Science promised readers a way of preparing for the uncertainties of a new order: large cities, democratic politics, unfamiliar forms of social organization, exotic foreign commodities, living with machines.

Books were at the cutting edge of the machine age, both in their production and as tools for intervening in public discussion. What made the lower cost of printed materials so exciting (or fearful) for contemporaries was that knowledge could be disseminated on a large scale. The new technologies of printing, publication, and distribution meant that a wide range of readers could access periodicals and books on a regular basis. The ramifications were not only intellectual, religious, and political, but also commercial. In the mid-1830s, medium-priced volumes (like most of those discussed here) overtook more expensive ones in the bookselling market for the first time. In context, the sales figures were unprecedented, reaching an audience an order of magnitude larger than had been usual for works on such subjects. At the same time, it is important to stress that the numbers of people who read such books remained modest in relation to the overall population. Take the *Constitution of Man,* the least expensive and best-selling of all these works, which up to 1860 had sold a cumulative 100,000 copies.[3] That is a remarkable number, even by today's standards, but still less than a third of 1 per cent of the population.

Historians have often claimed that attempts to diffuse knowledge in this period failed.[4] But this is to measure what actually occurred against the near-utopian hopes of projectors like Brougham. The noisy propaganda that surrounded the early Society for the Diffusion of Useful Knowledge has led to an assumption that the only significant potential audience for science was 'intelligent artisans' and other working people. It is certainly the case that the leaders of the 'march of intellect' targeted men like Croll and Claxton, in an effort to keep such men away from direct political action. This became clearest when the Society for the Diffusion of Useful Knowledge sponsored a pamphlet by Charles Knight on machine-breaking, which put the purveyors of useful knowledge of all kinds in direct opposition to those dedicated to the rights of labour. By that point, many radicals had come to see the movement to spread knowledge as nothing more than an attempt to distract the labouring classes from their true interests by plying them with stories about marsupials and the principles of mechanics. Writing in the weekly *Destructive* in 1834, the Irish radical Bronterre O'Brien said 'such knowledge is trash; the only knowledge which is any service to the working people is that which makes them more dissatisfied'.[5] Radical leaders warned against quiescence and complacency, while conservative Tories feared the result of the wide dissemination of knowledge would be revolution and atheism.

In any event, neither of these outcomes was realized. The market for science books proved to be primarily among the middle classes: shopkeepers, governesses, factory owners, lawyers, clerics, schoolteachers, and many others in commerce and the professions. A characteristic reader from the lower reaches of the middle class was Charles Tomlinson, whose family had fallen on hard times; among his first tasks as a young clerk was writing out in full Samuel Pepys's seventeenth-century shorthand diary. Tomlinson had attended evening lectures at the London Mechanics' Institute, saved his pennies for a weekly subscription to the Library of Useful Knowledge, and almost certainly read all of the works discussed here. His wife, Sarah Tomlinson (*née* Windsor), shared his scientific interests and wrote prolifically for the Society for Promoting Christian Knowledge. Tomlinson went on to produce technical books, including an authoritative two-volume encyclopaedia published in connection with the Great Exhibition of the Works of All Nations in Hyde Park in 1851.[6]

Just because readers of science were likely to have this kind of modest yet decidedly middle-class background does not mean that the cheapening of books somehow 'failed'. Rather than empowering or taming the revolutionary tendencies of the working class, the new literature of science was mainly important for developing the political order that emerged after the Reform Act was passed in 1832. As it became clear that the measure would do nothing for the great body of working people, the effectiveness of 'the march of intellect' and 'the schoolmaster is abroad' as rallying cries, uniting the movement for reform, quickly faded. By the mid-1830s the Society for the Diffusion of Useful Knowledge was no longer an important force in publishing, and its affairs were wound up after the failure of a biographical dictionary that had completed the letter 'A' after seven volumes.[7] The *Penny Magazine* and its many imitators were seen as politically compromised, or, what was worse, simply dull. To be successful, writing intended for a wide readership needed to combine science more explicitly with story-telling, entertainment, or a strong thread of theory.

Although the secular millennium of useful knowledge trumpeted by Brougham in 1825 never arrived, something else did. As it became clear that the worst nightmares of the conservatives were not going to be realized, a new consensus emerged among the governing classes.

Moderate Whigs and liberal Tories alike saw the Reform Act as a 'final' measure, with increased urban representation and the enfranchisement of a larger portion of the middle classes. The post-reform Whig government instituted measures limiting poor relief, improving sanitation, reducing the taxes that curbed the circulation of information, and extending Britain's overseas network of trade and influence. The British Association for the Advancement of Science, which met annually in different provincial cities throughout the Victorian period, was typical of many cross-party associations that developed from the early 1830s onwards on the basis of this new consensus.[8] Defeated and disappointed, the leaders of working-class opinion regrouped and reorganized, beginning campaigns for universal male suffrage, and the right for workers to strike and organize. While interest in science continued, the mainstream of opinion among working-class leaders focused on specific political objectives.

Although the utopian aspirations of the late 1820s and early 1830s quickly faded, they fed into a deeper and more lasting current of belief in progress. All of the books discussed here assumed that science, properly pursued, revealed the laws instituted by divine providence that underpinned the material and spiritual advance of civilization. As a Christian nation uniquely favoured by God, Britain had a special mission to spread enlightenment across the globe. Science was seen as the handmaiden of empire, not only through the work of exploration, mapping, and natural history surveys, but also in underwriting perceptions of European superiority and the virtues of governing those lower down in the scale of civilization. During the middle and later decades of the nineteenth century, the sciences played a vital role in developing new and more powerful visions of progress. Doctrines of the transmutation of species, once feared as threats to the Establishment, became central to defining a progressive order based on the laws of nature. The turning point occurred with the controversy over the anonymous *Vestiges of the Natural History of Creation*, published in

1844. *Vestiges* suggested that the main findings of contemporary science supported progressive development from the origins of the solar system to the creation of the human mind, and that these views need not be seen as leading to atheism or materialism. *Vestiges* was criticized on scientific grounds, but to many it was clear that this was the direction in which science was going.[9] For Charles Dickens, writing in the *Examiner* in 1848, the author of *Vestiges*—whom he almost certainly suspected was the journalist Robert Chambers—had succeeded in forging 'a reading public' for liberal scientific views of progress. Fifteen years after *Vestiges*, Charles Darwin's *On the Origin of Species* put the vital stamp of specialist authority on views that were already widely discussed by readers with a general interest in science. As Dickens had recognized, the people were already on Darwin's side.[10]

'Darwinism', as it was quickly termed, was only the most prominent of the unifying projects that enhanced the explanatory power of science and its global ambitions. In 1844 William Robert Grove's *On the Correlation of Physical Forces* argued for the reciprocal interdependence of heat, light, electricity, magnetism, and mechanical work.[11] This was part of a wider concern with the unity of powers in nature, an essential issue for the emerging industrial economy, which led during the second half of the century to mathematical formulations of the principle of the conservation of energy and the electromagnetic theory of light. Such unifying theories made books such as Somerville's *Connexion* and Herschel's *Preliminary Discourse* appear increasingly out of date. The new concepts grew out of practical concerns with empire—especially in relation to steam engines and telegraphy—and their success was to have profound consequences for global communications and industrial development. 'No one can contemplate the unexampled progress of science within the present century,' one author commented in 1852, 'without feeling that a new epoch has commenced in the history of our race.'[12]

The power increasingly accorded to science was symbolized by a widely circulated story about the seventeen-year-old Edward, Prince of Wales, who had been sent to Edinburgh by Queen Victoria and Prince Albert in 1859 to take science lessons from the chemist Lyon Playfair. Edward had never been much of a reader, but as a boy he had the benefit of a phrenological examination from Combe, as well as private tutoring by Charles Lyell and other scientific men. Standing near a cauldron of boiling lead, Playfair asked the future king if he had faith in science.

> 'Certainly,' replied the Prince.
>
> Playfair then carefully washed the Prince's hand with ammonia to get rid of any grease that might be on it.
>
> 'Will you now place your hand in this boiling metal, and ladle out a portion of it?' he said to his distinguished pupil.
>
> 'Do you tell me to do this?' asked the Prince.
>
> 'I do,' replied Playfair. The Prince instantly put his hand into the cauldron, and ladled out some of the boiling lead without sustaining any injury.

It was, apparently, 'a well-known scientific fact' that a perfectly clean hand could be placed in white-hot lead without any adverse effect. Repeated in sources ranging from the scientific weekly *Nature* (which had begun publication in 1869) to Sidney Lee's official life of King Edward VII, this story became a parable of the authority of science in the age of empire and industry.[13]

How did the new forms of natural science, still so controversial early in the nineteenth century, gain such authority? In part, this was achieved by the wide circulation of the books discussed in the preceding chapters. They appeared on the lists of respectable publishers, whose capital resources made possible the economies of cheap publishing by steam. Their authors were well known and genteel. They harnessed ideals of disinterestedness and the seeking out of truth. Books like the *Preliminary Discourse,* with carefully placed 'beauties' and stress on the moral probity of scientific investigation, made it more difficult to associate the sciences with irreligion and revolution. All of the works from these years drew explicit moral lessons from nature, in which the order of the physical universe was seen as a model for society and individual behaviour. Readers needed to be shown explicitly that science could be 'safe', leading 'from nature to nature's god'; but the passages that did this tended to be placed at the beginning and end of works. Natural theology, however useful in convincing readers that science need not lead to atheism and revolution, was becoming less significant as an apologetic strategy, as it was too distant from the central problems facing Christianity, such as the significance of Christ's atonement, the nature of original sin, and the evolution of Church doctrine. From this perspective, overt theological messages were necessary in the short term to clear the ground, as when *Consolations in Travel* and *Principles of Geology* confuted Scriptural literalists and Lamarckian materialists, or when *Connexion of the Physical Sciences* gave a divine lawmaker a place in the clockwork of the heavens.

The longer-term stabilizing effects of reading science were more subtle and pervasive than these framing materials would indicate. They were built into the texture of prose, embodied in the structure of arguments, and cemented by developing a relation of trust between author and reader. Readers could be drawn in by writing that assimilated narrative structures familiar from other literary genres, including novels, history, poetry, and periodical essays, at the same time placing these elements in larger contexts and imagined perspectives. As the poet Edward FitzGerald wrote in 1847, 'it is not the poetical imagination, but bare Science that every day more and more unrolls a greater Epic than the Iliad—the history of the World, the infinitudes of Space and Time! I never take up a book of Geology or Astronomy but this strikes me'.[14]

Most of all, however, these literary strategies appeared in a context that increasingly defined reading itself as a 'scientific' practice, in which relations between reader, author, and nature had the potential to avoid the clouding interventions of party, sect, self, or studied literary 'style'. Books as different as Herschel's *Preliminary Discourse* and Combe's *Constitution* were seen as conduct manuals, in which immersion in processes of induction and close observation guided readers to new forms of behaviour, both as individuals and in a hierarchy of collaboration. Science was seen to be wholesome, filling, and pure—in contrast to salacious novels, overtly political writings, or philosophies of nature from an earlier era. The new books of science demanded that readers follow arguments step by step, with close study and attention to detail. Above all, reading science encouraged the regulation of life according to reason and moderation, what became from the 1840s the guiding principles of political liberalism. It was what Herschel had pointed to as 'the gentle, but perfectly irresistible coercion' which habits of reading 'directed over the whole tenor of a man's character and conduct'.[15]

The new literature of scientific reflection had developed in the context of the 'march of intellect' as a peaceful alternative to using troops with fixed bayonets to end political unrest. It ended up, however, by contributing to the revolution in manners among the governing elite that emerged as a consequence of reform. The positive message of science was about everyday conduct, ordered hierarchies, and well-balanced minds. This was the point made by the liberal Whig weekly *Spectator*, in reviewing Lyell's *Principles*:

> The earth is an old reformer; her constitution has been subjected to innumerable changes; the signs of radical movements are to be detected everywhere, yet it is by no means easy to ascertain either the course or the causes of the revolutionary phenomena that so perpetually meet the eye of the inquirer. There are other investigations which more nearly affect our social happiness than the philosophy of geology, but perhaps there is none which in an indirect manner produce a more wholesome and beneficial effect upon the mind.... After the perusal of Mr. Lyell's volume, we confess to emotions of humility, to aspirations of the mind, to an

> elevation of thought, altogether foreign from the ordinary temper of worldly and busy men... So disposed mentally, the heart overflows with charity and compassion; vanity shrivels into nothingness; wrongs are forgotten, errors forgiven, prejudices fade away; the present is taken at its real value; virtue is tried by an eternal standard.[16]

The reader is transformed through the overall effect of engaging with scientific writing over an extended period, with its emphasis on scientific law, the uniformity of nature, and divine goodness. In the same vein, Somerville's *Connexion* and Herschel's *Preliminary Discourse* created among readers a shared perspective on truth, uniting calm sublimity with the precision of reasoned deliberation.

One of the fundamental legacies of the nineteenth century is a distinction between 'the literature of knowledge' and the 'literature of power', first broached by the essayist Thomas De Quincey in 1823. 'The function of the first,' De Quincey wrote in later elaborating his idea, 'is to *teach*; the function of the second is—to *move*.'[17] The literature of power draws upon the deepest feelings that make us human, motivating us towards engaged action. Like Milton's *Paradise Lost* and the great works of antiquity, the literature of power lasts forever. In contrast, the literature of knowledge is left behind by the onward march of progress. Even Newton's *Principia*, which De Quincey recognized as the 'very highest work that has ever existed in the literature of knowledge', turned out to have no more lasting literary qualities than a cookbook.[18] In many ways, from the perspective of the twenty-first century this distinction between power and knowledge may seem self-evident, from the texts chosen for close literary study to the preservation policies of libraries, where old science books are the first to be culled. The effect of the division is most pervasive in the asymmetrical approach we take to texts of different genres. Only naïve readers of *Jane Eyre* or *Pickwick Papers* assume that the author speaks directly through the voice of the narrator, but we do just that in reading the *Principles* or the *Preliminary Discourse*. Yet such works are engaging precisely because they do create a profound sense of a narrative presence—muted and shifting, earnest and ironic—which is all the more effective for being almost entirely invisible.

In distinguishing the 'literature of knowledge' from the 'literature of power', De Quincey was intervening in ongoing controversies about how the next generation—he addressed 'a young man whose education has been neglected'—could prepare for the brave new world of commerce and empire. Yet, for most contemporaries, his distinction did not hold, for as the revived slogan from Francis Bacon put it, knowledge was power. To read science was to enter imaginatively into the past, from the spectacle of condensing galaxies to the long records of life on earth. It offered vistas of the future, for although discoveries in chemistry, electricity, and other fields were only beginning to touch everyday life, they looked increasingly likely to change it significantly. Knowledge was considered to be useful not only in narrow economic terms, but also through its potential for enabling self-realization, emotional fulfilment, and social regeneration. The reading of science promised insight into the highest ideals of human conduct through reasoned debate, close observation, and dedication to truth. A great novel could move its readers, but so too could a profound work of science. In both cases, the literature of knowledge was the literature of power.

CHRONOLOGY

1819 Peterloo Massacre: cavalry charge into crowd demanding parliamentary reform in St Peter's Field in Manchester
Tory government passes the Six Acts, declaring that 'every meeting for radical reform is an overt act of treasonable conspiracy'
Birth of Princess Victoria

1820 Astronomical Society of London founded
Hans Christian Oersted announces unexpected relation between electricity and magnetism
Death of Joseph Banks; Humphry Davy becomes President of the Royal Society of London
Death of George III, accession of George IV
Queen Caroline, with public support and Henry Brougham as her advocate, successfully contests George IV's attempt to divorce her
Humphry Davy elected President of the Royal Society

1821 Richard Carlile's *Address to Men of Science*
Thomas De Quincey's *Confessions of an English Opium Eater*

1823 Mechanics' Institutes founded in England and Scotland

1824 Death of the poet Lord Byron
James Hogg's *The Private Memoirs and Confessions of a Justified Sinner*

1825 Henry Brougham's *Practical Observations upon the Education of the People*
Henry Brougham's inaugural discourse on education as Lord Rector of Glasgow University
Samuel Taylor Coleridge's *On the Constitution of Church and State*
William Hazlitt's *The Spirit of the Age*
Financial crash: failure of banks and difficulties for publishers

1826 London University (later University College London) established as counter to Oxford and Cambridge
Society for the Diffusion of Useful Knowledge founded
Early version of *Constitution of Man* circulated privately by George Combe

1827 Library of Useful Knowledge commences publication; Brougham asks Mary Somerville to contribute
Davies Gilbert elected as Royal Society President in succession to Humphry Davy

1828 Duke of Wellington becomes prime minister
George Combe's *Constitution of Man*
King's College launched at public meeting to oppose secular University of London
Board of Longitude closes
Death of natural philosopher William Hyde Wollaston
Publication of Sir Walter Scott's collected 'Magnum Opus' (1829–33) by Thomas Cadell

1829 Earl of Bridgewater's will announces provision for work (or works) 'On the Power, Wisdom, and Goodness of God as Manifested in the Creation'
Catholic Relief Bill passed
Death of the chemist Humphry Davy
'Signs of the Times' by Thomas Carlyle in *Edinburgh Review*
Newspaper disputes about management of the Royal Society of London
James South writes to *Times* about Humphry Davy's planned work on 'the decline of science'

1830 John Herschel's 'Sound', *Encyclopaedia Metropolitana*, discusses decline of science
Humphry Davy's *Consolations in Travel*
Dionysius Lardner's *Cabinet Cyclopaedia* (1830–44) published by Longman
Reports of speaking in tongues in Port Glasgow, Scotland
Charles Babbage's *Reflections on the Decline of Science in England*
Charles Lyell's *Principles of Geology*, volume I
Swing riots in agricultural districts
General election after death of George IV and accession of William IV
Liverpool and Manchester railway opens
David Brewster reviews Babbage's *Reflections*
First Whig government elected since 1783
Contested election for President of the Royal Society, John Herschel defeated by Duke of Sussex

1831 Thomas Love Peacock's *Crochet Castle* satirizes 'useful knowledge'
John Herschel's *Preliminary Discourse on the Study of Natural Philosophy*
Parliament dissolved during reform debate, election follows
First meeting of British Association for the Advancement of Science in York

Reform Bill defeated in House of Lords
Michael Faraday's experiments on electromagnetic induction
Riots in Derby, Nottingham, Bristol, and other cities
Mary Shelley's *Frankenstein* published by Coburn in revised, inexpensive edition
Mary Somerville's *Mechanism of the Heavens*
HMS *Beagle* sails with Charles Darwin on board

1832 Charles Lyell's *Principles of Geology*, volume II
Charles Babbage's *Economy of Machinery and Manufactures*
Penny Magazine of the Society for the Diffusion of Useful Knowledge
'Days of May': widespread riots and demonstrations with threats of revolution
Final passage of Reform Bill, removing worst abuses of 'old corruption'
Death of novelist and poet Sir Walter Scott
First election under provisions of Reform Bill

1833 Factory Act limits working hours of children
John Herschel's address to the Windsor and Eton Public Library subscribers
Charles Lyell's *Principles of Geology*, volume III
British Association for the Advancement of Science meets in Cambridge
Publication of first group of Bridgewater Treatises by Charles Bell, Thomas Chalmers, John Kidd, and William Whewell
Time ball installed at Greenwich Observatory
Fraser's Magazine publishes 'Sartor Resartus' by Thomas Carlyle in instalments (November 1833–August 1834)

1834 Richard Bentley publishes cheap collected edition of Jane Austen's novels
Mary Somerville's *On the Connexion of the Physical Sciences*
Failure of strike by tailors
Charles Lyell's *Principles of Geology* published by John Murray at reduced price
Death of Samuel Taylor Coleridge
Emancipation of slaves in the British Empire
New Poor Law ends outdoor relief for the destitute, offering minimal food and shelter only in workhouses
Election defeat for the Whigs; Robert Peel become prime minister but fails to form government

1835 Lord Melbourne becomes prime minister for Whigs (1835–41)
Margracia Loudon's *Philanthropic Economy*

	The socialist Robert Owen founds Association of All Classes of All Nations
	Cheap edition of George Combe, *Constitution of Man*, printed by W. & R. Chambers of Edinburgh
1836	Serial publication of Charles Dickens's *Pickwick Papers*
1837	Queen Victoria ascends to the throne

FURTHER READING

Desmond, Adrian. *The Politics of Evolution: Morphology, Medicine, and Reform in Radical London*. Chicago: University of Chicago Press, 1989. Revisionist account of the life sciences in the 1820s and 1830s, revealing radical challenges to the genteel world of science.

Gatrell, Vic. *City of Laughter: Sex and Satire in Eighteenth-Century London*. London: Atlantic Books, 2006. A vivid portrayal of changes in graphic satire, decorum, and manners through the early 1830s.

Hilton, Boyd. *A Mad, Bad, and Dangerous People? England, 1783–1846*. Oxford: Oxford University Press, 2006. Comprehensive survey by a leading historian, part of the New Oxford History of England. Especially good on politics, religion, and intellectual life.

Holmes, Richard. *The Age of Wonder: How the Romantic Generation Discovered the Beauty and Terror of Science*. London: HarperCollins, 2008. An entertaining and enlightening 'relay race of stories' by a leading literary biographer, focused on Joseph Banks, William Herschel, Humphry Davy, and others in the scientific generation that came to maturity in the second half of the eighteenth century.

Lightman, Bernard and Aileen Fyfe (eds). *Science in the Marketplace: Nineteenth-Century Sites and Experiences*. Chicago: University of Chicago Press, 2007. Excellent collection of scholarly essays dealing with a diverse range of audiences and sciences in nineteenth-century Britain.

Morrell, Jack and Arnold Thackray. *Gentlemen of Science: Early Years of the British Association for the Advancement of Science*. Oxford: Clarendon Press, 1981. Pioneering study of the social relations of science in the reform era, authoritative and meticulously detailed.

Morus, Iwan Rhys. *Frankenstein's Children: Electricity, Exhibition, and Experiment in Early Nineteenth-Century London*. Princeton: Princeton University Press, 1993. Accessible and insightful analysis of the world of experimental physics in relation to public life.

O'Connor, Ralph. *The Earth on Show: Fossils and the Poetics of Popular Science, 1802–1856*. Chicago: University of Chicago Press, 2007. Lively and sophisticated account of texts about antediluvian worlds and extinct monsters.

Pearce, Edward. *Reform! The Fight for the 1832 Reform Act*. London: Pimlico, 2004. Readable narrative of the parliamentary infighting that led to the passage of the Reform Act.

St Clair, William. *The Reading Nation in the Romantic Period*. Cambridge: Cambridge University Press, 2004. Major reinterpretation of literary publishing in the early nineteenth century.

Snyder, Laura J. *The Philosophical Breakfast Club: Four Remarkable Friends who Transformed Science and Changed the World*. New York: Broadway Books, 2011. Biographical treatment by a leading historian of philosophy of four major figures (Charles Babbage, John Herschel, William Jones, and William Whewell) and their ideas.

Winter, Alison. *Mesmerized: Powers of Mind in Victorian Britain*. Chicago: University of Chicago Press, 1998. Compelling account of the most controversial early Victorian theory of mind, and a model of how to study science in the making.

Yeo, Richard. *Defining Science: William Whewell, Natural Knowledge, and Public Debate in Early Victorian Britain*. Cambridge: Cambridge University Press, 1993. Important scholarly analysis of 'metascientific' writing in the 1830s.

ABBREVIATIONS

Works Discussed

CM George Combe, *The Constitution of Man Considered in Relation to External Objects* (Edinburgh: John Anderson; London: Longman, 1828), xii + 319 pp.
CPS Mary Somerville, *On the Connexion of the Physical Sciences* (London: John Murray, 1834), viii + 458 pp.
CT Humphry Davy, *Consolations in Travel, Or the Last Days of Philosopher* (London: John Murray, 1830), xii + 281 pp.
DS Charles Babbage, *Reflections on the Decline of Science in England, and on Some of Its Causes* (London: B. Fellowes and J. Booth, 1830), xvi + 228 pp.
PD John Frederick William Herschel, *A Preliminary Discourse on the Study of Natural Philosophy* (London: Longman, 1831), viii + 372 pp.
PG Charles Lyell, *Principles of Geology, Being an Attempt to Explain the Former Changes of the Earth's Surface, by Reference to Causes Now in Operation*, 3 vols (London: John Murray, 1830–33). Vol. 1 (1830, xv + 511 pp); Vol. 2 (1832, xii + 330 pp); Vol. 3 (1833, xxxii + 398 + 109 pp.
SR Thomas Carlyle, *Sartor Resartus*, ed. Kerry McSweeney and Peter Sabor (Oxford: Oxford University Press, 1987), xlii + 273 pp. The text of this edition is based on the original anonymous articles in monthly issues of *Fraser's Magazine* from November 1833 to August 1834.

All of the above books, other than *Sartor Resartus*, are available in well-produced facsimile editions from the Cambridge Library Collection published by Cambridge University Press.

Archives

BL British Library
CUL Cambridge University Library
NLS National Library of Scotland
RSL Royal Society of London

ENDNOTES

Introduction

1. 'Eton Subscription Library', *The Windsor and Eton Express* (2 February 1833). I am grateful to Alison Woodworth of Slough Library for tracking down this article. Herschel's address was published separately, distributed to the subscribers, and widely reprinted in contemporary periodicals. The most readily available edition is John Herschel, *Essays from the Edinburgh and Quarterly Reviews, with Addresses and Other Pieces* (London: Longman, 1857), 1–20.
2. Charles Knight, 'An Address to the Subscribers to the Windsor and Eton Public Library', *The Printing Machine; Or, Companion to the Library* (April 1834), 53–7.
3. The most important books of scientific reflection not discussed at length here, but published in the relevant period are the Bridgewater Treatises, which have been the subject by Jonathan Topham. His forthcoming book on their authorship, publication, and reception is eagerly awaited. The tradition continued in the Victorian era, with works by Michael Faraday (1839–55), William Whewell (1837 and 1840), Robert Chambers (1844), and Charles Darwin (1859), among many others.
4. Günther Buttmann, *The Shadow of the Telescope: A Biography of John Herschel* (New York: Charles Scribner's Sons, 1970), 113–16.
5. Margaret Oliphant, *Life of Edward Irving*, 2 vols (London: Hurst and Blackett, 1862), 1: 388–97. On contemporary millenarianism, see Boyd Hilton, *The Age of Atonement: The Influence of Evangelicalism on Social and Economic Thought, 1795–1865* (Oxford: Clarendon Press, 1988), and J. F. C. Harrison, *The Second Coming: Popular Millenarianism, 1780–1850* (London: Routledge, 1979).
6. On the new sense of history and the future in the early nineteenth century, see James Chandler, *England in 1819: The Politics of Literary Culture and the Case of Romantic Historicism* (Chicago: University of Chicago Press, 1998) and Reinhart Koselleck, *Futures Past: On the Semantics of Historical Time* (New York: Columbia University Press, 2004).

7. For these points and those in the following few paragraphs, see Andrew Cunningham and Perry Williams, 'De-centring the "Big Picture": *The Origins of Modern Science* and the Modern Origins of Science', *British Journal for the History of Science* 26 (1993), 407–32; Simon Schaffer, 'Scientific Discoveries and the End of Natural Philosophy', *Social Studies of Science* 16 (1986), 387–420; John Pickstone, 'Museological Science? The Place of the Analytical/Comparative in Nineteenth-Century Science, Technology, and Medicine', *History of Science* 32 (1994), 111–38; Jan Golinski, *Science as Public Culture: Chemistry and Enlightenment in Britain, 1760–1820* (Cambridge: Cambridge University Press, 1992); Adrian Desmond, *The Politics of Evolution: Morphology, Medicine, and Reform in Radical London* (Chicago: University of Chicago Press, 1989); and Roy Porter, *The Making of Geology: Earth Science in Britain, 1660–1815* (Cambridge: Cambridge University Press, 1977).
8. See Marilyn Butler, 'Introduction', in Mary Shelley, *Frankenstein or the Modern Prometheus: The 1818 Text* (Oxford: Oxford University Press, 1994).
9. Jonathan Topham has written extensively on the Bridgewater Treatises: see esp. his 'Science and Popular Education in the 1830s: The Role of the *Bridgewater Treatises*', *British Journal for the History of Science* 25 (1992), 397–430 and 'Beyond the "Common Context": The Production and Reading of the *Bridgewater Treatises*', *Isis* 89 (1998), 233–62.
10. On contemporary apocalyptic fears, see Hilton, *Age of Atonement*.
11. Alexander Gilchrist, *Life of William Blake, 'Pictor Ignotus'*, 2 vols (London: Macmillan, 1863), 1: 328.
12. David Vincent, *Literacy and Popular Culture* (Cambridge: Cambridge University Press, 1989), 22.
13. Henry Brougham, *Practical Observations upon the Education of the People, Addressed to the Working Classes and their Employers* (London: Longman, 1825), 32. My stress on the inclusiveness of the diffusionist programme is derived from Anne Secord, 'Botany on a Plate: Pleasure and the Power of Pictures in Promoting Early Nineteenth-Century Scientific Knowledge', *Isis* 93 (2002), 28–57, at pp. 29–30. For re-evaluations of Brougham's recasting of national politics, see William Hay, *The Whig Revival, 1808–1830* (Basingstoke: Palgrave Macmillan, 2004), and for its relations to science, see Joe Bord, *Science and Whig Manners: Science and Political Style in Britain, c. 1790–1850* (Basingstoke: Palgrave Macmillan, 2009).
14. Henry Brougham, *Inaugural Discourse of Henry Brougham, Esq., M. P., on Being Installed Lord Rector of the University of Glasgow* (Glasgow: Glasgow University Press, 1825), 47.
15. On printing and publishing, see Jonathan R. Topham, 'Scientific Publishing and the Reading of Science in Nineteenth-Century Britain: A Historiographical

Survey and Guide to Sources', *Studies in the History and Philosophy of Science* 31 (2000), 559–612; James Raven, *The Business of Books: Booksellers and the English Book Trade, 1450–1850* (New Haven and London: Yale University Press, 2007); and William St Clair, *The Reading Nation in the Romantic Period* (Cambridge: Cambridge University Press, 2004); and the relevant volumes of the *Cambridge History of the Book in Britain.*

16. Brougham, *Practical Observations*, 33. On Limbird, see Jonathan R. Topham, 'John Limbird, Thomas Byerley, and the Production of Cheap Periodicals in the 1820s', *Book History* 8 (2005), 75–106.
17. Charles Knight, *Passages of a Working Life during Half a Century: With a Prelude of Early Reminiscences*, 3 vols (London: Bradbury and Evans, 1864), 2: 66–7; also Valerie Gray, *Charles Knight: Educator, Publisher, Writer* (Aldershot: Ashgate, 2006).
18. John Gibson Lockhart, *Memoirs of the Life of Sir Walter Scott, Bart.*, 7 vols (Edinburgh: Robert Cadell, 1837–8), 6: 31.
19. Anne Secord, 'Introduction', in Gilbert White, *The Natural History of Selborne* (Oxford: Oxford University Press, 2013), xxvii; see also John Gibson Lockhart's comment on p. 245 of this edition.
20. Aileen Fyfe, *Steam-Powered Knowledge: William Chambers and the Business of Publishing, 1820–1860* (Chicago: University of Chicago Press, 2012).
21. Entry points into the literature on genre are provided by Clifford Siskin, *The Work of Writing: Literature and Social Change in Britain, 1700–1830* (Baltimore: Johns Hopkins University Press, 1998); Jon P. Klancher, *The Making of English Reading Audiences, 1790–1832* (Madison: University of Wisconsin Press, 1987); and Ralph O'Connor, 'Science as Literature', in *The Earth on Show: Fossils and the Poetics of Popular Science, 1802–1856* (Chicago: University of Chicago Press, 2007), 1–27, 217–61. For the diversity of fiction writing in this period, see Richard Maxwell and Katie Trumpener (eds), *The Cambridge Companion to Fiction in the Romantic Period* (Cambridge: Cambridge University Press, 2008).
22. 'The March of Intellect', *Morning Chronicle* (February 1828).
23. For closely related issues in France, see John Tresch, *The Romantic Machine: Utopian Science and Technology after Napoleon* (Chicago: University of Chicago Press, 2012), to which I am much indebted.
24. Although 'popularization' generally has this pejorative meaning today, a substantial historical literature explores the changing use of the word, along with related terms such as 'popular'. For discussion and further references, see Jonathan Topham et al., 'Historicizing "Popular Science"', *Isis* 100 (2009), 310–68; Bernard Lightman, *Victorian Popularizers of Science: Designing Nature for New Audiences* (Chicago: University of Chicago Press, 2007); Anne Secord, 'Science in the Pub: Artisan Botanists in Early Nineteenth-Century

Lancashire', *History of Science* 32 (1994), 383–408, esp. p. 299; and Roger Cooter and Stephen Pumpfrey, 'Separate Spheres and Public Places: Reflections on the History of Science Popularization and Science in Popular Culture', *History of Science* 32 (1994), 237–67.

25. Richard Yeo, *Defining Science: William Whewell, Natural Knowledge, and Public Debate in Early Victorian Britain* (Cambridge: Cambridge University Press, 1993).

Chapter 1

1. CT: 18.
2. CT: 25–6.
3. CT: 29.
4. CT: 43–4.
5. Acton Bell [Anne Brontë], *The Tenant of Wildfell Hall*, 3 vols (London: T. C. Newby, 1848) 1: 258–9.
6. Quoted in Clare Flaherty, 'A Recently Rediscovered Unpublished Manuscript: The Influence of Sir Humphry Davy on Anne Brontë', *Brontë Studies* 38 (2013), 30–41, at p. 38. For the geologist Gideon Mantell's related interest in the book, see Ralph O'Connor, 'Introduction', in *Victorian Science and Literature*, vol. 7: *Science as Romance* (London: Pickering & Chatto, 2012), xxiv.
7. John Murray, Ledger G, pp. 189, 231–9 *passim*, NLS ms 42727; Ledger E, pp. 31–2, 116, NLS ms 42730.
8. For a bibliography, see June Z. Fullmer, *Sir Humphry Davy's Published Works* (Cambridge, Massachusetts: Harvard University Press, 1969). *Consolations* did not sell well in the United States: there was only one separate edition, with a preface discounting the views expressed in the Vision.
9. CT: 170–1.
10. CT: 170.
11. C. Volney, *The Ruins, or a Survey of the Revolutions of Empires* (London: A. Seale, [1795]); Iain McCalman, *Radical Underworld: Prophets, Revolutionaries and Pornographers in London, 1795–1840* (Cambridge: Cambridge University Press, 1988), 20, 67.
12. Barbara B. Oberg et al. (eds), *The Papers of Thomas Jefferson*, 32 (Princeton: Princeton University Press, 2005), 441–42n.
13. William St Clair, *The Reading Nation in the Romantic Period* (Cambridge: Cambridge University Press, 2004), 317–22.
14. CT: 10–11.
15. For the motif of the eastern sage in contemporary literature, see Nigel Leask, *British Romantic Writers and the East: Anxieties of Empire* (Cambridge: Cambridge University Press, 1992).
16. CT: 15.

17. For biographical accounts of the author of *Consolations*, see esp. David Knight, *Humphry Davy: Science and Power* (Cambridge: Cambridge University Press, 1996) and June Z. Fullmer, *Young Humphry Davy: The Making of an Experimental Chemist* (Philadelphia: American Philosophical Society, 2000).
18. Henry Brougham, *Lives of Men of Letters and Science, who Flourished in the Time of George III* (London: Charles Knight, 1845), 466.
19. Jan Golinski, 'Humphry Davy's Sexual Chemistry', *Configurations* 7 (1999), 15–41, and also his 'Humphry Davy: The Experimental Self', *Eighteenth-Century Studies* 43 (2011), 15–28.
20. Author's collection.
21. CT: v–vi, ix.
22. I am indebted throughout this chapter to Jan Golinski's forthcoming work on the image of Davy as an author; see his 'Romantic Science: Humphry Davy's *Consolations in Travel*', Lindberg Lecture at the University of New Hampshire College of Liberal Arts, 11 Apr. 2013, <http://youtube/ZC9-5jtGlcc>.
23. CT: 103–5. On the ambiguity of the dream-vision and its significance for later scientific speculations, see O'Connor, 'Introduction', *Science as Romance*, xxiii–xxv.
24. CT: vi.
25. CT: 92–3.
26. CT: 224–5.
27. On the mentoring genre, see Adrian J. Wallbank, *Dialogue, Didacticism and the Genres of Dispute: The Literary Dialogue in an Age of Revolution*. (London: Pickering & Chatto, 2012).
28. CT: 135, 137.
29. CT: 149.
30. CT: 150.
31. CT: 219.
32. CT: 280–1.
33. CT: 254.
34. Jan Golinski, *Science as Public Culture: Chemistry and Enlightenment in Britain, 1760–1820* (Cambridge: Cambridge University Press, 1992), 188–235.
35. [Humphry Davy], *Salmonia: or, Days of Fly Fishing. In a Series of Conversations, with Some Account of the Habits of Fish belonging to the Genus Salmo* (London: John Murray, 1828), 183–4.
36. 'Salmonia', *Blackwood's Edinburgh Magazine* 24 (August 1828), 248–72, at pp. 248, 250, 254, 258.
37. W., 'Value of Religion', *Wesleyan-Methodist Magazine* 8 (October 1829), 681.
38. Jane Davy to John Murray II, 1 November 1828, NLS ms 40320, ff. 34–5.

39. Sarah Harriet Burney to Charlotte Barrett, 1 September 1831, in Lorna J. Clark (ed.), *The Letters of Sarah Harriet Burney* (Athens, Georgia: University of Georgia Press, 1997), 355.
40. *Edinburgh Literary Journal* (13 March 1830), 153–4, at p. 154; *Literary Gazette* (27 February 1830), 134–5, at p. 135.
41. 'Monthly View of New Publications', *La Belle Assemblee: or Court and Fashionable Magazine* 11 (April 1830), 165.
42. [Edward Cox], 'Life and Writings of Sir Humphry Davy', *Dublin Review* 2 (April 1837), 437–74, at pp. 473–4.
43. Humphry Davy to Jane Davy, 15 March 1827, in John Davy, *Fragmentary Remains, Literary and Scientific, of Sir Humphry Davy, Bart.* (London: John Churchill, 1858), 279–80.
44. 'Daviana', *Catholic Magazine* 3 (1833), 375–82, quoting CT: 163.
45. See esp. [Davy], *Salmonia*, 182–3.
46. Humphry Davy to Jane Davy, 25 April 1827, in Davy, *Fragmentary Remains*, 283–4; see also [Davy], *Salmonia*, 182–3.
47. Charles Tomlinson, '"Blunders of a Big-Wig." Anonymous', *Notes and Queries*, ser. 8, 7 (9 February 1895), 109–10, at p. 110.
48. Brougham, *Lives*, 466; Golinski, 'Humphry Davy's Sexual Chemistry', 26–7.
49. Richard Carlile, *Address to Men of Science* (London: Richard Carlile, 1821), 21.
50. John Murray, *A Handbook for Travellers in Southern Germany*, 2nd ed. (London: John Murray, 1840), 131, 209, 210, 211, 339, 351, 357.
51. Author's collection. See also Gary J. Tee, 'Relics of Davy and Faraday in New Zealand', *Notes and Records of the Royal Society of London* 52 (1998), 93–102.
52. Brian Stapleford, 'Science Fiction before the Genre', in Edward James and Farah Mendlesohn (eds), *The Cambridge Companion to Science Fiction* (Cambridge: Cambridge University Press, 2003), 15–31, at p. 21.

Chapter 2

1. CT: 226.
2. Francis Lunn to Charles Babbage, 28 November [1830], cited in Jack Morrell and Arnold Thackray, *Gentlemen of Science: Early Years of the British Association for the Advancement of Science* (Oxford: Clarendon Press, 1981), 49.
3. Simon Schaffer, 'Babbage's Intelligence: Calculating Engines and the Factory System', *Critical Inquiry* 21 (1994), 203–27.
4. For these debates in relation to political economy, see esp. Maxine Berg, *The Machinery Question and the Making of Political Economy, 1815–1848* (Cambridge: Cambridge University Press, 1980).

5. See esp. Schaffer, 'Babbage's Intelligence'.
6. 'Censor', 'To the Editor of the Times' (3 December 1829), 3.
7. Thomas Young to the Admiralty, 10 January 1829, quoted in Alexander Wood, *Thomas Young: Natural Philosopher, 1773–1829* (Cambridge: Cambridge University Press, 1954), 311.
8. David Philip Miller, 'Between Hostile Camps: Sir Humphry Davy's Presidency of the Royal Society of London, 1820–1827', *British Journal for the History of Science* 16 (1983), 1–43, esp. pp. 3–8.
9. William J. Ashworth, 'The Calculating Eye: Baily, Herschel, Babbage and the Business of Astronomy', *British Journal for the History of Science* 27 (1994), 409–41; for the 'grand amateurs', Alan Chapman, *The Victorian Amateur Astronomer: Independent Astronomical Research in Britain, 1820–1920* (London: John Wiley, 1998).
10. John Barrow, *Sketches of the Royal Society Club* (London: John Murray, 1849), 10.
11. Miller, 'Between Hostile Camps'.
12. Adrian Desmond, *The Politics of Evolution: Morphology, Medicine, and Reform in Radical London* (Chicago: University of Chicago Press, 1989).
13. Sophie Waring, 'Thomas Young and the Board of Longitude', abstract of paper delivered at British Society for the History of Science annual meeting, 2011.
14. James South, 'Royal Society', *The Times* (11 December 1829), 4.
15. DS: x.
16. [David Brewster], 'Decline of Science in England', *Edinburgh Journal of Science*, n.s., 5 (October 1831), 334–58, at p. 334.
17. 'Socius', 'To the Editor of the Times', *The Times* (10 December 1829), 3; 'Paul Pry', 'Royal Society', *The Times* (9 December 1829), 4; [Augustus Bozzi Granville], 'One of the 680 F.R.S.', 'Royal Society', *The Times* (2 December 1830), 3; 'Job Hater', *The Times* (10 December 1829), 3; 'Censor', 'To the Editor of the Times', *The Times* 3 December 1829), 3; 'A Constant Reader', 'Royal Society and Arundel Mss.', *The Times* (11 December 1829), 4.
18. DS: ix–xi.
19. Mary Louise Gleason, *The Royal Society of London, 1827–1847* (New York: Garland Publishing, 1991), 183–4.
20. Charles Babbage to John Herschel, 19 March 1830, quoted in Gleason, *Royal Society*, 182–3.
21. DS: xi.
22. Charles Babbage to John Herschel, 19 March 1830, quoted in Gleason, *Royal Society*, 182–3.
23. DS: 48–9. Robert Brown was expelled early in 1830; see 'The Medico-Botanical Society and the Ex-Director', *Spectator* (23 January 1830), 51. The Zoological

Society also erupted into controversy: see Adrian Desmond, 'The Making of Institutional Zoology in London, 1822–1836', *History of Science* 23 (1985), 153–85, 223–50.

24. *The Times* (8 May 1830), 2.
25. James South, *Charges Against the President and Councils of the Royal Society* (London: Richard Clay, 1830); [Augustus Bozzi Granville], *Science Without a Head: or the Royal Society Dissected by One of the 687 F.R.S.—s* (London: T. Ridgway, 1830).
26. David Brewster to Charles Babbage, 12 February 1830, in Jack Morrell and Arnold Thackray (eds), *Gentlemen of Science: Early Correspondence of the British Association for the Advancement of Science* (London: Royal Historical Society, 1984), 24–5.
27. [William Whewell], 'Cambridge Transactions: Science of the English Universities', *British Critic* 9 (January 1831), 71–90 and 'Lyell—*Principles of Geology*', *British Critic* 9 (January 1831), 180–206, at pp. 181–2.
28. The best account of the relation between the declinist controversy and the British Association for the Advancement of Science is Morrell and Thackray, *Gentlemen of Science: Early Years*, 35–94.
29. DS: 77–100, at p. 78.
30. DS: 97.
31. Edward Sabine, 'Notices Occasioned by the Perusal of a Late Publication by Mr Babbage', *Philosophical Magazine* n.s., 8 (July 1830), 45–50.
32. DS: x.
33. 'Finsbury Election. Mr Babbage', *The Times* (23 November 1832), 1.
34. Harriet Martineau, *Autobiography*, 3 vols (London: Smith, Elder & Co., 1877), 1: 354.
35. Journal entry for 8 January 1832, K. M. Lyell, *Life Letters and Journals of Sir Charles Lyell, Bart.*, 2 vols (London: John Murray, 1881), 1: 363.
36. 'Critical Notices', *New Monthly Review* 59 (February 1831), 31.
37. DS: 36.
38. DS: 41.
39. DS: 43–4.
40. DS: 174–83.
41. DS: 169.
42. DS: 182.
43. *Decline* makes it clear that in this case only the positive side of the balance sheet is being discussed; Wollaston's secrecy had been condemned earlier (DS: 133–4) and Davy's political machinations as Royal Society President became the subject of controversy many years later, after the publication of Babbage's *Passages from the Life of a Philosopher* (London: Longman, 1864), 144–6.

44. Charles Babbage, *On the Economy of Machinery and Manufactures* (London: Charles Knight, 1832); on the publishing history of this book, and Babbage's attitude towards print, see Adrian Johns, 'The Identity Engine: Printing and Publishing at the Beginning of the Knowledge Economy', in Lissa Roberts, Simon Schaffer, and Peter Dear (eds), *The Mindful Hand: Inquiry and Invention from the Late Renaissance to Early Industrialisation* (Amsterdam: Koninklijke Nederlandse Akademie van Wetenschappen, 2007), 403–28.
45. James Millingen, *Some Remarks on the State of Learning and the Fine-Arts in Great Britain, on the Deficiency of Public Institutions, and the Necessity of a Better System for the Improvement of Knowledge and Taste* (London: J. Rodwell, 1831), 13, 22–9.
46. William Wilkins, *A Letter to Lord Viscount Goderich, On the Patronage of the Arts by the English Government* (London: unpublished, 1832); published in *Library of the Fine Arts* 3 (April–June 1832), 291–307, 367–78, 472–83.
47. Edward Bulwer-Lytton, *England and the English*, 2 vols (London: Richard Bentley, 1833); for some of the other general discussions, see [Richard Henry Horne], *Exposition of the False Medium and Barriers Excluding Men of Genius from the Public* (London: Effingham Wilson, 1833) and William Swainson, *Preliminary Discourse on the Study of Natural History* (London: Longmans, 1839), 296–450.
48. 'State of Science and of Learned Societies in England', *Asiatic Journal* n.s., 2 (June 1830), 120–5.
49. 'F.R.S.' [William Martin Leake], *Thoughts on the Degradation of Science in England* (London: John Rodwell, 1847), 25–6.
50. 'New Borough of Finsbury', *Morning Chronicle* (10 December 1832), 2. On the campaigns, see Anthony Hyman, *Charles Babbage: Pioneer of the Computer* (Oxford: Oxford University Press, 1984), 75–87.
51. 'Mr Babbage and the Interests of Science', *Mechanics' Magazine* 17 (8 December 1832), 159; for the election, see Hyman, *Charles Babbage*, 82–7.
52. 'New Borough of Finsbury', *Morning Chronicle* 10 December 1832), 2.
53. 'Timothy Tickler' [John Gibson Lockhart], 'The Purists of Politics. A Poem', *Blackwood's Edinburgh Magazine* 32 (October 1832), 413–30, at p. 22, n. 18.
54. William Wordsworth to William Rowan Hamilton, 24 January 1831, in Alan G. Hill and Ernest de Selincourt (eds), *The Letters of William and Dorothy Wordsworth*, 2nd ed. (Oxford: Clarendon Press, 1979), 5 (2): 364.
55. 'A Foreigner' [Gerard Moll], *On the Alleged Decline of Science in England* (London: T. & T. Boosey, 1831), 20. See also Nathan Reingold, 'Babbage and Moll on the State of Science in Great Britain: A Note on a Document', *British Journal for the History of Science* 4 (1968), 58–64.
56. Charles Babbage, *The Exposition of 1851; or, Views of the Industry, the Science, and the Government of England*, 2nd ed. (London: John Murray, 1851), 6, 11.

57. Thomas Ledger, 'Phrenological Examination', 17 June 1852, BL Add. Ms 37195, ff. 193–6. Thanks to Simon Schaffer for alerting my attention to this. A page from another consultation is reproduced in Alison Winter, *Mesmerized: Powers of Mind in Victorian Britain* (Chicago: University of Chicago Press, 1998), 151. See also Thomas Leger, *The Magnetoscope: A Philosophical Essay on the Magnetoid Characteristics of Elementary Principles, and their Relations to the Organisation of Man* (London: Bailliere, 1852), and Victor Horsley, 'Description of the Brain of Mr Charles Babbage, F.R.S.', *Philosophical Transactions of the Royal Society of London* 200 (1909), 117–31.
58. DS: 212.
59. Babbage, *Economy of Machinery*, 320.
60. Simon Schaffer, 'Babbage's Dancer and the Impresarios of Mechanism', in Francis Spufford and Jenny Uglow (eds), *Cultural Babbage: Time, Technology and Invention* (London: Faber, 1996), 53–60; see also Charles Babbage, *The Ninth Bridgewater Treatise: A Fragment* (London: John Murray, 1837).

Chapter 3

1. For introductions to the large and sophisticated literature in this field, see Laura J. Snyder, *Reforming Philosophy: A Victorian Debate on Science and Society* (Chicago: University of Chicago Press, 2006), and Richard Yeo, *Defining Science: William Whewell, Natural Knowledge, and Public Debate in Early Victorian Britain* (Cambridge: Cambridge University Press, 1993).
2. James Raven, *The Business of Books: Booksellers and the English Book Trade 1450–1850* (New Haven: Yale University Press, 2007); Richard B. Sher, *The Enlightenment and the Book: Scottish Authors and their Publishers in Eighteenth-Century Britain, Ireland, and America* (Chicago: University of Chicago Press, 2006), esp. 27–9.
3. The plan for the work is frequently bound into copies of individual volumes; see also 'Dr Lardner's Cyclopaedia', *The Times* (4 December 1829), 3, and J. N. Hays, 'The Rise and Fall of Dionysius Lardner', *Annals of Science* 38 (1981), 527–42. On Longman's importance for science publishing, see Jonathan Topham, 'Scientific Publishing and the Reading of Science in Early Nineteenth-Century Britain: An Historiographical Survey and Guide to Sources', *Studies in History and Philosophy of Science* 31A (2000), 559–612.
4. John Herschel to Dionysius Lardner, 1 August 1828, RSL: Herschel mss 11: 103; and John Herschel to Dionysius Lardner, 28 February 1829, summarized in Michael J. Crowe et al. (eds), *A Calendar of the Correspondence of Sir John Herschel* (Cambridge: Cambridge University Press, 1998).
5. Dionysius Lardner to John Herschel, 19 January 1830, RSL: Herschel mss 11: 115.
6. John Herschel to David Brewster, 25 November 1831 and 8 December 1831, RSL: Herschel mss, 25: 3.1 and 28.

7. Cited in Rosemary Ashton, *Victorian Bloomsbury* (New Haven and London: Yale University Press, 2012), 89.
8. John Herschel, 'Sound', in *Encyclopaedia Metropolitana*, 4 (*Mixed Sciences*, Vol. 2), (London: John Joseph Griffin, part issued 3 February 1830): 747–824, 810. This article is misdated in much of the literature, including the generally helpful account in Susan Faye Cannon, *Science in Culture: The Early Victorian Period* (Folkestone: Dawson, 1978), 181.
9. John Herschel to Charles Babbage, [5–18 March 1830], RSL: Herschel ms 2: 245, quoted in Jack Morrell and Arnold Thackray, *Gentlemen of Science: Early Years of the British Association for the Advancement of Science* (Oxford: Oxford University Press, 1981), 48.
10. John Herschel to John F. Daniell, 24 November 1831, RSL: Herschel mss, 25.2: 29.
11. Marie Boas Hall, *All Scientists Now: The Royal Society in the Nineteenth Century* (Cambridge: Cambridge University Press, 1984), 57–62; on Herschel's views on decline, Cannon, *Science in Culture*, 181–6.
12. John Herschel to William Henry Fitton, 18 October 1830, quoted in Cannon, *Science in Culture*, 183.
13. PD: 93.
14. John F. W. Herschel, 'An Address to the Subscribers to the Windsor and Eton Public Library and Reading Room', in *Essays from the Edinburgh and Quarterly Reviews, with Addresses and other Pieces* (London: Longman, 1857), 1–20, at p. 13.
15. John Herschel to William Somerville, 16 September 1831, RSL: Herschel ms 16: 393.
16. PD: 16–17.
17. PD: viii, 3–4, 8.
18. PD: 83.
19. PD: 84.
20. PD: 98.
21. PD: 103, 99.
22. PD: 103.
23. PD: 190–1.
24. PD: 193–4.
25. PD: 194–5.
26. PD: 348–53.
27. PD: 15–16.
28. Quoted in Morrell and Thackray, *Gentlemen of Science: Early Years*, 292.
29. PD: 74.
30. PD: 4.

31. James A. Secord, *Victorian Sensation: The Extraordinary Publication, Reception and Secret Authorship of* Vestiges of the Natural History of Creation (Chicago: University of Chicago Press, 2000), 162–3, and Alice Walters, 'Conversation Pieces: Science and Politeness in Eighteenth-Century England', *History of Science* 35 (1997), 121–54.
32. 'Review of Herschel's Discourse on the Study of Natural Philosophy', *Wesleyan-Methodist Magazine* 10 (November 1831), 762–8; (December 1831), 842–9, at pp. 762, 763; and *Monthly Magazine* n.s., 11 (February 1831), 225.
33. Leah Price, *The Anthology and the Rise of the Novel: From Richardson to George Eliot* (Cambridge: Cambridge University Press, 2000).
34. [William Whewell], 'Modern Science—Inductive Philosophy', *Quarterly Review* 45 (July 1831), 374–407, at p. 406.
35. [John A. Carlyle], 'Natural Philosophy', *Fraser's Magazine* 3 (July 1831), 698–702, at p. 700.
36. 'Mr Herschel's Natural Philosophy', *The Times* (28 February 1831), 5.
37. Walter F. Cannon, 'John Herschel and the Idea of Science', *Journal of the History of Ideas* 22 (1961), 215–39, at pp. 218–19. On Darwin as a reader of Herschel, see Michael Ruse, 'Darwin's Debt to Philosophy: An Examination of the Influence of the Philosophical Ideas of John F. W. Herschel and William Whewell on the Development of Charles Darwin's Theory of Evolution', *Studies in the History and Philosophy of Science* 6 (1975), 159–81.
38. Charles Darwin, *On the Origin of Species by Means of Natural Selection* (London: John Murray, 1859), 1.
39. Charles Darwin, 'Recollections of the Development of My Mind and Character', in Darwin, *Evolutionary Writings*, ed. James A. Secord (Oxford: Oxford University Press, 2008), 381–2.
40. Charles Darwin to William Darwin Fox, [15 February 1831], in Frederick Burkhardt et al. (eds), *The Correspondence of Charles Darwin* (Cambridge: Cambridge University Press, 1985), 1: 118.
41. James Mackintosh, *History of England* (London: Longman, 1831), 2: 114–15n; also James Mackintosh to Frances Allen, 8 March 1831, in Robert James Mackintosh (ed.), *Memoirs of the Life of the Right Honourable Sir James Mackintosh*, 2 vols (London: Edward Moxon, 1835), 2: 480.
42. PD: 192, 267, 359, 360; see William Jackson, *The Four Ages; With Essays on Various Subjects* (London: Cadell & Davies, 1798).
43. [John Stuart Mill], 'Herschel's *Preliminary Discourse*', *Examiner* (20 March 1831), 179–80.
44. John Stuart Mill, *A System of Logic, Ratiocinative and Inductive, Being a Connected View of the Principles of Evidence, and the Methods of Scientific Investigation*, 2 vols (London: John W. Parker, 1838), 1: 490.

45. [Whewell], 'Modern Science—Inductive Philosophy', 382.
46. William Whewell, *The Philosophy of the Inductive Sciences*, 2 vols (London: John W. Parker, 1840), 1: cxiii; [William Whewell], 'On the Connexion of the Physical Sciences', *Quarterly Review* 51 (March 1834), 54–68, at p. 59.
47. Laura J. Snyder, *The Philosophical Breakfast Club: Four Remarkable Friends who Transformed Science and Changed the World* (New York: Broadway Books, 2011), 297–8.
48. Quoted from *Science-Gossip* n.s., 1 (1894), 242–3, at p. 243, in Melinda Clare Baldwin, *Making* Nature: *The History of a Scientific Journal* (Chicago: University of Chicago Press, forthcoming).
49. Paul White, *Thomas Huxley: Making the 'Man of Science'* (Cambridge: Cambridge University Press, 2003), Ruth Barton, '"Men of Science": Language, Identity and Professionalization in the Mid-Victorian Scientific Community', *History of Science* 41 (2003), 73–119, and Baldwin, *Making* Nature. See also Sydney Ross, '*Scientist*: The Story of a Word', *Annals of Science* 18 (1962), 65–85.
50. [Whewell], 'On the Connexion', 65.

Chapter 4

1. *Mechanics' Magazine* 20 (29 March 1834), 442–7, at p. 447.
2. James Clerk Maxwell, 'Grove's "Correlation of Physical Forces"', *Nature* 10 (20 August 1874), 302–4, at p. 303. The special problems of dealing with mathematical sciences are discussed in Alice Jenkins, 'Genre and Geometry: Victorian Mathematics and the Study of Literature and Science', in Ben Marsden, Hazel Hutchinson, and Ralph O'Connor (eds), *Uncommon Contexts: Encounters between Science and Literature, 1800–1914* (London: Pickering & Chatto, 2013), 111–23.
3. Jonathan Topham, 'Science and Popular Education in the 1830s: The Role of the *Bridgewater Treatises*', *British Journal of the History of Science* 25 (1992), 397–430, at pp. 413–14; for a closer look at one treatise, see James Sumner, *Brewing Science, Technology and Print, 1700–1880* (London: Pickering & Chatto, 2013), 159–61.
4. Henry Brougham, *Objects, Advantages and Pleasures of Science* (London: Baldwin and Cradock, 1827), 40.
5. Henry Brougham to William Somerville, 27 March 1827, in Martha Somerville (ed.), *Personal Recollections, from Early Life to Old Age, of Mary Somerville* (London: John Murray, 1873), 161–2.
6. Brougham to W. Somerville, 27 March 1827, in Somerville, *Personal Recollections*, 161–2.

7. Brougham to W. Somerville, 27 March 1827, in Somerville, *Personal Recollections*, 161.
8. Henry Brougham, *Practical Observations upon the Education of the People, Addressed to the Working Classes and their Employers* (London: Longman, 1825), 10.
9. [Mary Somerville], 'A Lady', 'Prize Question 310', in Thomas Leybourn (ed.), *New Series of the Mathematical Repository* 3.1 (London: W. Glendinning, 1811), Mathematical Questions, 151–3. Reprinted in Mary Somerville, *Collected Works of Mary Somerville*, ed. James A. Secord, 9 vols (Bristol: Thoemmes Press, 2004), vol. 1.
10. Somerville, *Personal Recollections*, 80. These books are now in the Library of Girton College, Cambridge.
11. Somerville, *Personal Recollections*, 130. For this setting, see Elizabeth Chambers Patterson, *Mary Somerville and the Cultivation of Science, 1815–1840* (Boston: Martinus Nijhoff, 1983), 35–53.
12. Charles Caldwell, *Autobiography* (Philadelphia: Lippincott, 1855), 380–3.
13. James Secord, 'How Scientific Conversation became Shop Talk', in Aileen Fyfe and Bernard Lightman (eds), *Science in the Marketplace: Nineteenth-Century Sites and Experiences* (Chicago: University of Chicago Press, 2007), 23–59.
14. Henry Kater to H.-C. Oersted, 17 February 1826, quoted in Elizabeth Chambers Patterson, *Mary Somerville and the Cultivation of Science, 1815–1840* (Boston: Martinus Nijhoff, 1983), 47.
15. Mary Somerville to John Herschel, [March 1830], quoted in Patterson, *Mary Somerville*, 75.
16. R. K. Webb, *The British Working Class Reader, 1790–1848: Literacy and Social Tension* (London: George Allen & Unwin, 1955), 69.
17. Augustus De Morgan, *The Differential and Integral Calculus* (London: Baldwin and Cradock, 1836–42), 5n.
18. 'Useful Knowledge', *Westminster Review* 14 (April 1831), 384; for similar views of the *Mechanism of the Heavens*, see the review in the *Athenaeum* (21 January 1832), 43–4.
19. George Peacock to Mary Somerville, in Somerville, *Personal Recollections*, 172. On the publication history, see James A. Secord, 'Introduction', in Somerville, *Collected Works*, 2: ix–xvi, at pp. xii–xiv.
20. Joanna Baillie to Mary Somerville, 1 February 1832, in Martha Somerville (ed.), *Personal Recollections, from Early Life to Old Age, of Mary Somerville* (London: John Murray, 1873), 206.
21. William Somerville to Janet Somerville Pringle Elliot, 29 February 1832, in Patterson, *Mary Somerville*, 88. Patterson assumes, like Kathryn A. Neeley, *Mary Somerville: Science, Illumination, and the Female Mind* (Cambridge: Cambridge University Press, 2001) that the 'Preliminary Dissertation' was separately

published, but in fact this is not the case; see Secord, 'Introduction', in Somerville, *Collected Works,* 2: ix–xvi, at p. xiv.

22. Patterson, *Mary Somerville,* 119. The printer was William Clowes.
23. Scott Bennett, 'John Murray's Family Library and the Cheapening of Books in Early Nineteenth Century Britain', *Studies in Bibliography* 29 (1976), 139–66, at p. 141.
24. [William Cooke Taylor], Review of *On the Connexion of the Physical Sciences, Athenaeum* (15 March 1834), 202–3.
25. Dorothy McMillan (ed.), *Queen of Science: Personal Recollections of Mary Somerville* (Edinburgh: Canongate Classics, 2001), 74. Somerville's attitudes towards women's rights are best discussed in Robyn Arianrhod, *Seduced by Logic: Émilie Du Châtelet, Mary Somerville and the Newtonian Revolution* (Oxford: Oxford University Press, 2012), 164–8.
26. On this issue, see Susan Faye Cannon, *Science in Culture: The Early Victorian Period* (Folkestone: Dawson, 1978), 111–36; Iwan Rhys Morus, *When Physics became King* (Chicago: University of Chicago Press, 2005).
27. CPS: v.
28. James Clerk Maxwell, 'Grove's "Correlation of Physical Forces"', *Nature* 10 (20 August 1874), 302–4, at p. 303.
29. CPS: 5.
30. 'Character of Modern Works on Physical Sciences', *British Critic* 16 (July 1834), 123–32, at p. 132.
31. Kathryn A. Neeley, *Mary Somerville: Science, Illumination, and the Female Mind* (Cambridge: Cambridge University Press, 2001), 101–29.
32. CPS: 171.
33. CPS: 172.
34. CPS: 404.
35. Maria Edgeworth to Mary Somerville, 31 May 1832, in Somerville, *Personal Recollections,* 204.
36. 'Character of Modern Works on Physical Sciences', *British Critic* 16 (July 1834), 123–32, at p. 132; Harriet Martineau, *Biographical Sketches. 1852–1875,* new ed. (London: Macmillan, 1875), 496. For an informative account of Somerville's status in relation to genres of popular writing, see Claire Brock, 'The Public Worth of Mary Somerville', *British Journal for the History of Science* 39 (2006), 255–72.
37. Maria Edgeworth to Mary Somerville, 31 May 1832, in Somerville, *Personal Recollections,* 204.
38. CPS: 3–4.
39. CPS: 194.

40. CPS: 251, 285; for the role of the ether, see Alice Jenkins, *Space and the 'March of Mind': Literature and the Physical Sciences in Britain, 1815–1850* (Oxford: Oxford University Press, 2007), 176–207.
41. CPS: 413–14.
42. Somerville, *Personal Recollections,* 28.
43. Mary Somerville to Thomas Somerville, 8 August 1819, Bodleian Library, Ms Dep. c.360 MSFP-44.
44. Mary Somerville to Thomas Somerville, 2 March 1821, Bodleian Library, Ms Dep. 360 MSFP-44.
45. 'Character of Modern Works on Physical Sciences', *British Critic* 16 (July 1834), 123–32, at p. 130. On this issue generally, see Aileen Fyfe, *Science and Salvation: Evangelicals and Popular Science Publishing in Victorian Britain* (Chicago: University of Chicago Press, 2004).
46. [William Whewell], 'On the Connexion of the Sciences', *Quarterly Review* 51 (March 1834), 54–68, at p. 66.
47. [David Brewster], 'Mrs Somerville on the Physical Sciences', *Edinburgh Review* 59 (April 1834), 154–71, at pp. 170–1.
48. Woronzow Greig to Mary Somerville, 9 April 1835, in Patterson, *Mary Somerville,* 163. On the civil list pension, see Brock, 'Public Worth of Mary Somerville'.
49. [Thomas Galloway], 'Mrs Somerville's *Mechanism of the Heavens*', *Edinburgh Review* 55 (April 1832), 1–25, at p. 1; Review of *On the Connexion of the Physical Sciences, The Printing Machine* (19 April 1834), 78–80, at p. 78; [John Herschel], 'Mrs. Somerville's *Mechanism of the Heavens*', *Quarterly Review* 47 (July 1832), 537–59, at p. 548.
50. [Francis Sylvester Mahoney], 'The Days of Erasmus', *Fraser's Magazine* 11 (May 1835), 559–75, at p. 560.
51. Londa Schiebinger, *Nature's Body: Women in the Origins of Modern Science* (Boston: Beacon Press, 1989).
52. [Whewell], 'On the Connexion', 65.
53. Mary Somerville to Martha Somerville, 3 January [1835], in Patterson, *Mary Somerville,* 179; Brock, 'Public Worth of Mary Somerville', 262.
54. On this episode, see Secord, 'General Introduction', in Somerville, *Collected Works,* 1: xv–xxxix, at pp. xxxv–xxxvii, and Patterson, *Mary Somerville,* 142–3.
55. CPS: 413.

Chapter 5

1. Charles Lyell to Roderick I. Murchison, 17 January 1829, in Leonard G. Wilson, *Charles Lyell: The Years to 1841: The Revolution in Geology* (New Haven and London: Yale University Press, 1972), 256. An earlier version of

the present chapter appeared as the introduction to Charles Lyell, *Principles of Geology* (London: Penguin Books, 1997), ix–xliii.

2. Charles Southwell, 'A View from Bristol Gaol', *Oracle of Reason* 1 (19 February 1842), 78–9.
3. Roy Porter, *The Making of Geology: Earth Science in Britain, 1660–1815* (Cambridge: Cambridge University Press, 1977), and Rachel Laudan, *From Mineralogy to Geology: The Foundations of a Science, 1650–1830* (Chicago: University of Chicago Press, 1987). On the cultural significance of geology, especially in literature, see Ralph O'Connor, *The Earth on Show: Fossils and the Poetics of Popular Science, 1802–1856* (Chicago: University of Chicago Press, 2007), and Noah Heringman, *Romantic Rocks, Aesthetic Geology* (Ithaca, New York: Cornell University Press, 2004).
4. Roy Porter, 'Charles Lyell: The Public and Private Faces of Science', *Janus* 69 (1982), 29–50.
5. Charles Lyell to William Whewell, 20 February 1831, in Wilson, *Charles Lyell*, 308.
6. Adrian Desmond, *Politics of Evolution: Medicine, Morphology, and Reform in Radical London* (Chicago: University of Chicago Press, 1989), 328.
7. Wilson, *Charles Lyell*, 320–2.
8. Charles Lyell to Mary Horner, 17 November 1831, in K. M. Lyell (ed.), *Life Letters and Journals of Sir Charles Lyell*, 2 vols (London: John Murray, 1881), 1: 352–3.
9. Charles Lyell to George Poulett Scrope, 14 June 1830, in Lyell, *Life*, 1: 271.
10. Charles Lyell to John Murray, 30 September 1830, NLS Add. ms 40726, ff. 46–7.
11. The book was printed by William Clowes. For publication figures from the Murray Archive, see Stuart A. Baldwin, *Charles Lyell: A Brief Bibliography* (Wickham Bishops, Witham: Baldwin's Scientific Books, 2013), 3.
12. Timothy L. Alborn, 'The Business of Induction: Industry and Genius in the Language of British Scientific Reform, 1820–1840', *History of Science* 34 (1996), 91–121; for the older networks, see Anne Secord, 'Corresponding Interests: Artisans and Gentlemen in Nineteenth-Century Natural History', *British Journal for the History of Science* 27 (1994), 383–408, and David Allen, *The Naturalist in Britain: A Social History* (Harmondsworth: Penguin, 1976).
13. Charles Lyell, *A Second Visit to the United States of America*, 2 vols (London: John Murray, 1850), 2: 317; John M. Clarke, *James Hall of Albany, Geologist and Paleontologist, 1811–1898* (Albany, 1921), 35.
14. Laudan, *From Mineralogy to Geology*; Rachel Laudan, 'The Role of Methodology in Lyell's Science', *Studies in History and Philosophy of Science* 13 (1982), 215–49; see also PD: 146–7.

15. PG1: 6.
16. Adam Sedgwick, 'Address to the Geological Society', *Proceedings of the Geological Society of London* 1 (1831), 281–316, at p. 307.
17. Charles Lyell to Gideon Mantell, 15 February 1830, in Lyell, *Life*, 1: 262.
18. PG1: 2.
19. PG1: 123. For a close reading of this passage and of the *Principles* more generally, see O'Connor, *Earth on Show*, 163–87.
20. Martin J. S. Rudwick, 'Caricature as a Source for the History of Science: De la Beche's Anti-Lyellian Sketches of 1831', *Isis* 66 (1975), 534–60; on Lyell and the reconstruction of earth history, see Dov Ospovat, 'Lyell's Theory of Climate', *Journal of the History of Biology* 10 (1977), 317–39, and esp. Martin J. S. Rudwick, *Worlds Before Adam: The Reconstruction of Geohistory in the Age of Reform* (Chicago: University of Chicago Press, 2008).
21. Stephen Jay Gould, *Time's Arrow, Time's Cycle* (Cambridge, Massachusetts: Harvard University Press, 1987), 99–189. For a subtle analysis of the use of narrative in *Principles*, see Adelene Buckland, *Novel Science: Fiction and the Invention of Nineteenth-Century Geology* (Chicago: University of Chicago Press, 2013).
22. Rudwick, *Worlds Before Adam*, 302–3.
23. Charles Lyell, journal entry for 7 April 1832, in Lyell, *Life*, 1: 377.
24. Dugald Stewart, *Elements of the Philosophy of the Human Mind* (London: A. Strahan and T. Cadell, 1792), sect. 1, ch. 3.
25. *Scotsman* (25 September 1830), 1. For a list of reviews, see Charles Lyell, *Principles of Geology*, ed. James A. Secord (London: Penguin Books, 1997), 459–61.
26. Martin J. S. Rudwick, 'Bibliography of Lyell's Sources', in Charles Lyell, *Principles of Geology*, 3 vols (Chicago: University of Chicago Press, 1990–1), 3: 113–60.
27. Dennis Dean, *James Hutton and the History of Geology* (Ithaca, New York: Cornell University Press, 1992), 239.
28. PG1: 81–2.
29. PG1: 161.
30. Charles Lyell to Charles Lyell, Sr, 31 October 1816, in Lyell, *Life*, 1: 38–9. Pietro Corsi, 'The Heritage of Dugald Stewart: Oxford Philosophy and the Method of Political Economy', *Nuncius* 2 (1987), 89–144.
31. John Herschel to Charles Lyell, 20 February 1836, in Walter F. Cannon, 'The Impact of Uniformitarianism', *Proceedings of the American Philosophical Society* 105 (1961), 301–14, at p. 305.
32. William Whewell, *History of the Inductive Sciences*, 3 vols (London: John W. Parker, 1837), 3: 448.

33. *1832 British Association Report*, 406; *Monthly Magazine* n.s., 10 (December 1830), 701; *Literary Gazette* (14 August 1830), 526; *New Monthly Magazine* (1 June 1832), 241.
34. James Bonwick, cited in John Burnett (ed.), *Destiny Obscure: Autobiographies of Childhood, Education and Family from the 1820s to the 1920s* (London: Allen Lane, 1982), 173.
35. Charles Lyell to George Ticknor, 1850, in Lyell, *Life*, 2: 168–9.
36. Sharon Turner, *The Sacred History of the World, Attempted to be Philosophically Considered, in a Series of Letters to a Son*, 8th ed., 3 vols (London: Longman, 1848), 1: x. On Scriptural geology, see Ralph O'Connor, 'Young-Earth Creationists in Early Nineteenth-Century Britain? Towards a Reassessment of "Scriptural Geology"', *History of Science* 45 (2007), 357–403.
37. Charles Lyell to George Poulett Scrope, 14 June 1830, in Lyell, *Life*, 1: 268.
38. Nicolaas Rupke, *The Great Chain of History: William Buckland and the English School of Geology* (Oxford: Clarendon, 1983).
39. Charles Lyell to George Poulett Scrope, 14 June 1830, and Lyell to Gideon Mantell, 29 December 1827, in Lyell, *Life*, 1: 271, 174.
40. PG3: x–xi. See Rudwick, *Worlds Before Adam*, 276–82.
41. John McPhee, *Basin and Range* (New York: Farrar, Straus and Giroux, 1981).
42. PG1: 63.
43. Charles Lyell to George Poulett Scrope, 14 June 1830, in Lyell, *Life*, 1: 268–71.
44. Charles Lyell to Gideon Mantell, 29 December 1827, in Lyell, *Life*, 1: 173–4.
45. Charles Lyell to Eleanor Lyell, 26 February 1830, in Lyell, *Life*, 1: 263.
46. Charles Lyell to Mary Horner, 14 February 1832, in Wilson, *Charles Lyell*, 343.
47. Charles Lyell to Mary Horner, 14 February 1832, in Wilson, *Charles Lyell*, 344.
48. Charles Lyell to John Murray, 31 August 1834, NLS ms 40726, ff. 132–3. He also thought of writing an 'original & popular' work for the Family Library; see Charles Lyell to John Murray, 22 March 1830, NLS ms 40726, ff. 38–9.
49. Charles Lyell to John Murray, 1 October 1833, NLS ms 40726, ff. 107–8.
50. Basil Hall to Leonard Horner, 7 September 1833, in Lyell, *Life*, 2: 466.
51. Charles Lyell to Eleanor Lyell, 11 May 1830, in Lyell, *Life*, 1: 267.
52. Quoted in Dennis Dean, '"Through Science to Despair": Geology and the Victorians', in James Paradis and Thomas Postlewait (eds), *Victorian Science and Victorian Values* (New York: New York Academy of Sciences, 1985), 111–36, at p. 115.
53. *Metropolitan Magazine* 10 (May 1834): 'Literature', 4–5.
54. *British Critic* 15 (April 1834), 334–63, at p. 363; *New Monthly Magazine* (1 June 1832), 241–2, at p. 21; *Presbyterian Review and Religious Journal* 2 (July 1832), 329–45, at p. 339; *Eclectic Review* n.s., 6 (July 1831), 75–81, at p. 79.
55. Sedgwick, 'Address', 313.
56. PG3: 385.

57. PG2: 18f.
58. Desmond, *Politics of Evolution*; James A. Secord, 'Edinburgh Lamarckians: Robert Jameson and Robert E. Grant', *Journal of the History of Biology* 24 (1991), 1–18.
59. *Monthly Review* (March 1832), 352–71, at p. 353.
60. PG2: 2.
61. CT: 149–50.
62. PG1: 144–66.
63. Sedgwick, 'Address', 305.
64. Charles Lyell to John Herschel, 1 June 1836, in Lyell, *Life*, 2: 467; Charles Lyell to Edward Copleston, 28 March 1831, in Wilson, *Charles Lyell*, 310.
65. Charles Lyell to Edward Copleston, 28 March 1831, in Wilson, *Charles Lyell*, 310; for more on this incident, see Martin J. S. Rudwick, 'Charles Lyell, F.R.S. (1797–1875) and his London Lectures on Geology, 1832–33', *Notes and Records of the Royal Society of London* 29 (1975), 231–63.
66. Charles Lyell to John Herschel, 1 June 1836, in Lyell, *Life*, 2: 467; Charles Lyell to Edward Copleston, 28 March 1831, in Wilson, *Charles Lyell*, 310.
67. Robert Jameson to John Murray, 18 February 1832, quoted in Janet Browne, *Charles Darwin: Voyaging* (London: Jonathan Cape, 1995), 188.
68. *Gentleman's Magazine* 102 (January 1832), pt I, 43–7, at p. 45; *Spectator* (14 January 1832), 39; *Literary Gazette* (28 January 1832), 49.
69. Leonard G. Wilson (ed.), *Sir Charles Lyell's Scientific Journals on the Species Question* (New Haven and London: Yale University Press, 1970); Michael Bartholomew, 'Lyell and Evolution', *British Journal for the History of Science* 6 (1973), 261–303.
70. Charles Lyell to Gideon Mantell, 2 March 1827, in Lyell, *Life*, 1: 163, and Buckland, *Novel Science*, 66–7.
71. Charles Lyell to Caroline Lyell, 19 October 1830, in Lyell, *Life*, 1: 308.
72. On the long-term effects, see Rudwick, *Worlds Before Adam*.
73. D. R. Stoddart, 'Darwin, Lyell, and the Geological Significance of Coral Reefs', *British Journal for the History of Science* 9 (1976), 199–218; Browne, *Charles Darwin*, 186–90; James A. Secord, 'The Discovery of a Vocation: Darwin's Early Geology', *British Journal for the History of Science* 24 (1991), 133–57; Sandra Herbert, *Charles Darwin, Geologist* (Ithaca, New York: Cornell University Press, 2005).

Chapter 6

1. Especially useful on phrenology in Britain are Roger Cooter, *The Cultural Meaning of Popular Science: Phrenology and the Organization of Consent in Nineteenth-Century Britain* (Cambridge: Cambridge University Press, 1984); John

van Wyhe, *Phrenology and the Origins of Victorian Scientific Naturalism* (Aldershot: Ashgate, 2004) and David Stack, *Queen Victoria's Skull* (London: Hambleton Continuum, 2008), which is a biography of George Combe.

2. On Loudon's philanthropic writings, novels, and career, see Sarah Richardson, *The Political Worlds of Women: Gender and Politics in Nineteenth Century Britain* (New York: Routledge, 2013), esp. pp. 5–15 and 73–5, and also her 'Women, Philanthropy, and Imperialism in Early Nineteenth-Century Britain', in Helen Gilbert and Chris Tiffin (eds), *Burden or Benefit? Imperial Benevolence and its Legacies* (Bloomington and Indianapolis: Indiana University Press, 2008), 90–102.
3. James Kennedy to George Combe, 5 November 1826, NLS ms 7217, ff. 132–3.
4. James Kennedy to George Combe, 8 October 1835, NLS ms 7235, ff. 131–4.
5. Letter from Margracia Loudon to James Kennedy, quoted in Kennedy to George Combe, 8 October 1835, NLS ms 7235, ff. 131–4.
6. Letter from Margracia Loudon to James Kennedy, quoted in Kennedy to George Combe, 8 October 1835, NLS ms 7235, ff. 131–4.
7. Margracia Loudon to George Combe, 7 January 1836, NLS ms 7239, ff. 143–5.
8. James Kennedy to George Combe, 8 October 1835, NLS ms 7235, ff. 131–4.
9. Margracia Loudon, *The Light of Mental Science; Being an Essay on Moral Training* (London: Smith, Elder, 1845), 3.
10. George Combe to Margracia Loudon, 24 December 1845, NLS ms 7390, ff. 243–5.
11. CM: 140.
12. Cooter, *Cultural Meaning*, 224–55, at p. 245.
13. Margracia Loudon, 'Remarks on the *Constitution of Man*', *Monthly Repository* 10 (1836), 153–8. Other than in Richardson's work, this review has hitherto been incorrectly attributed to the natural history writer Jane Loudon.
14. George Combe to Margracia Loudon, 29 February 1836, NLS ms 7386, ff. 507–8.
15. Charles Gibbon, *The Life of George Combe, Author of 'The Constitution of Man'*, 2 vols (London: Macmillan, 1878), 2: 14.
16. Gibbon, *Life of George Combe*, esp. 1: 94–6, and Stack, *Queen Victoria's Skull*.
17. George Combe, *A System of Phrenology*, 5th ed., 2 vols (Edinburgh: Maclachlan, 1843), 1: vii.
18. James A. Secord, 'Science, Technology and Mathematics', in David McKitterick (ed.), *The Cambridge History of the Book in Britain, Volume VI, 1830–1914* (Cambridge: Cambridge University Press, 2009), 443–74.
19. William Godwin, *Thoughts on Man, his Nature, Productions, and Discoveries. Interspersed with some Particulars Respecting the Author* (London: Effingham Wilson, 1831).

20. There is no adequate study of Hope's book and its reception, but see Roger Scruton, 'Hope's Philosophical Excursus', in David Watkin and Philip Hewat-Jaboor (eds), *Thomas Hope: Regency Designer* (New Haven and London: Yale University Press, 2008), 243–7.
21. Adrian Desmond, *The Politics of Evolution Morphology, Medicine and Reform in Radical London* (Chicago: University of Chicago Press, 1989), 117–21.
22. CM: 289.
23. Cooter, *Cultural Meaning*, 189; van Wyhe, *Phrenology*, 156, also reproduces the title page.
24. On *Constitution*'s early history, see van Wyhe, *Phrenology*, 96–164 and Stack, *Queen Victoria's Skull*, 47–93.
25. George Combe, *Testimonials on Behalf of George Combe as a Candidate for the Chair of Logic in the University of Edinburgh* (Edinburgh: John Anderson, 1936), 72; for Neill, see van Wyhe, *Phrenology*, 160.
26. Van Wyhe, *Phrenology*, 132–4.
27. Aileen Fyfe, *Steam-Powered Knowledge: William Chambers and the Business of Publishing, 1820–1860* (Chicago: University of Chicago Press, 2012).
28. Combe, *Testimonials on Behalf of George Combe*, 55.
29. Robert Chambers to George Combe, 25 November 1835, NLS ms 7234, ff. 140–1.
30. William Pearson, *The Constitution of Man Considered in Relation to External Objects* (New York: William Pearson, 1835).
31. On the Henderson editions, see van Wyhe, *Phrenology*, 133–4, and Fyfe, *Steam-Powered Knowledge*, 71–2. That Chambers, more than Combe, was the moving force behind the cheap editions is indicated by Robert Chambers to George Combe, 24 September 1835, NLS ms 7234, ff. 138–9, and William Fraser to Robert Cox, 10 June 1835, NLS ms 7235, ff. 35–6.
32. Richardson, 'Women, Philanthropy, and Imperialism', 97.
33. CM: 1.
34. George Combe, *The Constitution of Man Considered in Relation to External Objects*, 2nd ed. (Edinburgh: John Anderson, 1835), 1.
35. Combe, *Constitution*, 25.
36. Frederick Douglass, *Life and Times of Frederick Douglass, Written by Himself* (Hartford, Connecticut: Park, 1882), 276.
37. George Combe to Gustav von Struve, 21 October 1845, in Gibbon, *Life of George Combe*, 2: 375–83.
38. Douglass, *Life and Times*, 276.
39. Frances Anne Kemble, *Records of a Girlhood*, 2 vols (London: Richard Bentley, 1878), 1: 245–7; Gibbon, *Life of George Combe*, 2: 241–2.
40. [Adam Sedgwick], 'Natural History of Creation', *Edinburgh Review* 82 (July 1845), 1–85, at p. 13.

41. [Robert Chambers], 'The Easily Convinced', *Chambers's Edinburgh Journal* (2 July 1842), 185–6.
42. [John Playfair], 'Transactions of the Geological Society', *Edinburgh Review* 19 (November 1811), 207–29, at p. 207.
43. Gibbon, *Life of George Combe*, 2: 329.
44. Gibbon, *Life of George Combe*, 2: 328.
45. Gibbon, *Life of George Combe*, 2: 327.
46. Gibbon, *Life of George Combe*, 2: 327.
47. See Edward Copeland, *The Silver Fork Novel: Fashionable Fiction in the Age of Reform* (Cambridge: Cambridge University Press, 2012); on satires of fashion, see Diana Donald, *Followers of Fashion: Graphic Satires from the Georgian Period* (London: Haywood Gallery, 2002), and Diana Donald, *The Age of Caricature: Satirical Prints in the Reign of George III* (New Haven: Yale University Press, 1996). Sharrona Pearl, *About Faces: Physiognomy in Nineteenth-Century Britain* (Cambridge, Massachusetts: Harvard University Press, 2010) discusses the relation between outward appearance and inner character more generally.
48. The attribution is based on the copy in the Bodleian Library, Opie BB118.
49. Catherine Sinclair, *Modern Accomplishments, or the March of Intellect* (Edinburgh: Waugh & Innes, 1836), vi.
50. Sinclair, *Modern Accomplishments*, 6–10, 220, 221.
51. David Russell, '"Our Debt to Lamb": The Romantic Essay and the Emergence of Tact', *ELH* 79 (2012), 179–209.

Chapter 7

1. SR: 241. For orientations, see G. B. Tennyson, Sartor *called* Resartus: *The Genesis, Structure, and Style of Thomas Carlyle's First Major Work* (Princeton: Princeton University Press, 1965), and Rodger L. Tarr, 'Introduction', in Thomas Carlyle, *Sartor Resartus: The Life and Opinions of Herr Teufelsdröckh in Three Books*, ed. Roger L. Tarr (Berkeley and Los Angeles: University of California Press, 2000), xxi–xciv. For a subtle reading of the ironies of the text see Anne K. Mellor, 'Carlyle's *Sartor Resartus*: A Self-Consuming Artifact', in *English Romantic Irony* (Cambridge, Massachusetts: Harvard University Press, 1980), 109–34. For a fine recent discussion in relation to the sciences, see Charlotte Sleigh, *Literature and Science* (Basingstoke: Palgrave Macmillan, 2011), 86–90.
2. SR: 45.
3. For an excellent analysis of the serial context, see Mark Parker, *Literary Magazines and British Romanticism* (Cambridge: Cambridge University Press, 2000), 157–81. On *Fraser's*, see Miriam M. H. Thrall, *Rebellious Fraser's: Nol Yorke's Magazine in the Days of Maginn, Thackeray, and Carlyle* (New York:

Columbia University Press, 1934) and David E. Latané, *William Maginn and the British Press: A Critical Biography* (Farnham: Ashgate, 2013). I have used quotation marks around 'Sartor' because it was not at first a book.

4. SR: 3.
5. Carlyle, *Sartor Resartus*, 239.
6. SR: 3–4.
7. SR: 8–9.
8. SR: 26.
9. SR: 34–6.
10. SR: 87.
11. SR: 112.
12. SR: 124, 118.
13. For contemporary descriptions, see 'Time Ball on the Greenwich Observatory', *Arcana of Science and Art* 7 (1834), 13–14, and 'Neptune', 'The Time-Ball at Greenwich', *Nautical Magazine* 4 (October 1835), 584–6. 'Sartor' had been largely completed in 1831, so the innovation at Greenwich is relevant to the reception of the work, rather than its composition. The national adoption of Greenwich time as a standard took at least half a century: see David Rooney and James Nye, '"Greenwich Observatory Time for the Public Benefit": Standard Time and Victorian Networks of Regulation', *British Journal for the History of Science* 42 (2009), 5–30.
14. SR: 44.
15. SR: 194–5.
16. Job 38: 4.
17. SR: 203.
18. SR: 206.
19. SR: 217, and Iwan Rhys Morus, *Frankenstein's Children: Electricity, Exhibition, and Experiment in Early Nineteenth-Century London* (Princeton: Princeton University Press, 1993).
20. James A. Secord, 'Scrapbook Science: Composite Caricatures in Late Georgian London', in Ann B. Shteir and Bernard Lightman (eds), *Figuring it Out: Science, Gender, and Visual Culture* (Hanover, New Hampshire: Dartmouth College Press, 2006), 164–91, at pp. 180–2. For dandies, see Diana Donald, *Followers of Fashion: Graphic Satires from the Georgian Period* (London: Haywood Gallery, 2002).
21. SR: 220. John B. Lamb, '"Spiritual Enfranchisement": *Sartor Resartus* and the Politics of *Bildung*', *Studies in Philology* 107 (2010), 259–82 discusses *Sartor* and working-class politics. For the strike of 1834 and its aftermath, see E. P. Thompson, *The Making of the English Working Class* (London: Penguin Books, 1980), 279–85.

22. SR: 51.
23. Carlyle may have seen these words above the Tolbooth, a former prison on the Canongate in Edinburgh, before it was demolished in 1817. See Carlyle, *Sartor Resartus*, 445.
24. The following discussion is indebted to Carlisle Moore, 'Carlyle: Mathematics and "Mathesis"', in K. J. Fielding and Rodger L. Tarr (eds), *Carlyle Past and Present: A Collection of New Essays* (London: Vision Press, 1976), 61–95. For Carlyle's later scientific interests, see John M. Ulrich, 'Thomas Carlyle, Richard Owen, and the Paleontological Articulation of the Past', *Journal of Victorian Culture* 11 (2006), 30–56, and Rebecca Stott, 'Thomas Carlyle and the Crowd: Revolution, Geology and the Convulsive "Nature" of Time', *Journal of Victorian Culture* 4 (1999), 1–24.
25. Thomas Carlyle to Alexander Carlyle, 9 August 1821, in C. R. Sanders and K. J. Fielding (eds), *The Collected Letters of Thomas and Jane Welsh Carlyle* 1 (Durham, North Carolina: Duke University Press, 1970), 376–7; hereafter, *Collected Letters*. On these works, see Moore, 'Carlyle: Mathematics and "Mathesis"', 61–95.
26. Augustus De Morgan, *A Budget of Paradoxes* (London: Longmans, 1872), 499.
27. Thomas Carlyle to Robert Mitchell, 25 May 1818, in *Collected Letters* 1 (1970), 128.
28. [Thomas Carlyle], 'Signs of the Times', *Edinburgh Review* 59 (June 1829), 439–59, at p. 445.
29. See Lawrence Poston, 'Millites and Millenarians: The Context of Carlyle's "Signs of the Times"', *Victorian Studies* 26 (1983), 381–406.
30. SR: 53.
31. [Carlyle], 'Signs of the Times', 448, 449.
32. [Carlyle], 'Signs of the Times', 443.
33. Thomas Carlyle to John A. Carlyle, 30 April 1835, in *Collected Letters* 8 (1981), 100–6.
34. Thomas Carlyle to Jane Welsh Carlyle, 11 July 1843, in *Collected Letters* 16 (1990), 260.
35. Thomas Carlyle to John A. Carlyle, 21 October 1831, in *Collected Letters* 6 (1977), 30.
36. 4 February 1839, in *Collected Letters* 11 (1985), 19, and Thomas Carlyle to John A. Carlyle, 24 November 1840, in *Collected Letters* 12 (1985), 335–6.
37. [William Maginn], 'Babbage on Machinery and Manufactures', *Fraser's Magazine* 8 (August 1833), 167–75, at p. 172.
38. Thomas Carlyle to Macvey Napier, 8 October 1831, in *Collected Letters* 6 (1977), 13.
39. [Carlyle], 'Signs of the Times', 443–4.

40. [Robert Mudie], *Babylon the Great: Or, Men and Things in the British Capital*, 2nd ed., 2 vols (London: Henry Colburn, 1828), 2: 228.
41. 'Miss Martineau, Fraser's Magazine, and the Conservative Press', *Liverpool Mercury* (29 November 1833).
42. 'On the March of Intellect, and Universal Education', *Fraser's Magazine* 2 (September 1830), 161–70, at p. 162.
43. 'On the March of Intellect', 163–4.
44. [Francis Mahoney], 'The Days of Erasmus', *Fraser's Magazine* 11 (May 1835), 559–79, at p. 560.
45. Thomas Love Peacock, *Nightmare Abbey* (London: T. Hookham, 1818), 30, 111.
46. *The Sun* (1 April 1834), reprinted in D. J. Trela and Rodger L. Tarr (eds), *The Critical Response to Thomas Carlyle's Major Works* (Westport, Connecticut: Greenwood Press, 1997), 13.
47. SR: 81–2.
48. J. Beete Jukes to Andrew Ramsay, 14 July 1850, in C. A. Jukes-Browne (ed.), *Letters and Extracts from the Addresses and Occasional Writings of J. Beete Jukes* (London: Chapman & Hall, 1871), 447.
49. 'George Acorn' (pseudonym), *One of the Multitude* (London: W. Heinemann, 1911), 193. For such readers of *Sartor*, see Jonathan Rose, *The Intellectual Life of the British Working Classes* (New Haven: Yale University Press, 2001), 41–8.
50. Frank M. Turner, 'Victorian Scientific Naturalism and Thomas Carlyle', in *Contesting Cultural Authority: Essays in Victorian Intellectual Life* (Cambridge: Cambridge University Press, 1993), 142, quoting from T. H. Huxley, 'Professor Tyndall', *The Nineteenth Century* 35 (1894), 1–11, at p. 3.
51. Ruth Barton, 'John Tyndall: Pantheist: A Rereading of the Belfast Address', *Osiris*, 2nd ser., 3 (1987), 111–34, quoting from a letter of 1850 from John Tyndall to Thomas Archer Hirst.
52. SR: 4, 5.
53. SR: 16.
54. SR: 16, and note on 251; also SR: 4. The 'mistake' is also identified in Carlyle, *Sartor Resartus* (2000), note on p. 255.

Epilogue

1. James Campbell Iron, *Autobiographical Sketch of James Croll...with Memoir of his Life and Work* (London: Edward Stanford, 1896), 12–13; Timothy Claxton, *Memoir of a Mechanic. Being a Sketch of the Life of Timothy Claxton, Written by Himself* (Boston: George W. Light, 1839), 36.
2. William Wordsworth, 'Preface to *Lyrical Ballads* (1802)', in *The Major Works*, ed. Stephen Gill (Oxford: Oxford University Press, 1984), 595–620.

3. John van Wyhe, *Phrenology and the Origins of Victorian Scientific Naturalism* (Aldershot: Ashgate, 2004), 128.
4. This is particularly apparent in the literature on mechanics' institutes and the Society for the Diffusion of Useful Knowledge, in works going back to R. K. Webb, *The British Working Class Reader, 1790–1848: Literacy and Social Tension* (London: George Allen & Unwin, 1955). There are notable exceptions, esp. Rosemary Ashton, *Victorian Bloomsbury* (New Haven and London: Yale University Press, 2012), 58–81 and Valerie Gray, *Charles Knight: Educator, Publisher, Writer* (Aldershot: Ashgate, 2006).
5. Bronterre O'Brien, *Destructive* (7 June 1834), quoted in Patricia Hollis, *Class and Conflict in Nineteenth Century England: 1815–1850* (London: Routledge, 1973), 336. The literature on radical opposition to the plans to diffuse knowledge is very extensive.
6. See the article in the *Oxford Dictionary of National Biography*, and Charles Tomlinson, '"Blunders of a Big-Wig". Anonymous', *Notes and Queries*, ser. 8, 7 (9 February 1895), 109–10.
7. Ashton, *Victorian Bloomsbury*, 77–80. Technically these were 'half-volumes', each of several hundred pages.
8. For the wider political context, see esp. Boyd Hilton, *A Mad, Bad, and Dangerous People? England, 1783–1846* (Oxford: Oxford University Press, 2006); Peter Mandler, *Aristocratic Government in the Age of Reform: Whigs and Liberals, 1830–1852* (Oxford: Clarendon Press, 1990); and Jonathan Parry, *The Rise and Fall of Liberal Government in Victorian Britain* (New Haven and London: Yale University Press, 1996).
9. James A. Secord, *Victorian Sensation: The Extraordinary Publication, Reception and Secret Authorship of* Vestiges of the Natural History of Creation (Chicago: University of Chicago Press, 2000).
10. [Charles Dickens], 'Review: *The Poetry of Science, or Studies of the Physical Phenomena of Nature*, by Robert Hunt', in Michael Slater (ed.), *Dickens's Journalism*, 4 vols (London: J. M. Dent, 1996), 129–34, at p. 131, and esp. the annotated edition of this text in Ralph O'Connor (ed.), *Victorian Science and Literature*, vol. 7: *Science as Romance* (London: Pickering & Chatto, 2012), 13–20.
11. Iwan Rhys Morus, 'Correlation and Control: William Robert Grove and the Construction of a New Philosophy of Scientific Reform', *Studies in the History and Philosophy of Science* 22 (1991), 589–621.
12. Michael Angelo Garvey, *The Silent Revolution: Or the Future Effects of Steam and Electricity upon the Condition of Mankind* (London: William and Frederick G. Cash, 1852), 1–2.
13. Wemyss Reid, *Memoir and Correspondence of Lyon Playfair* (London: Cassell, 1899), 201. The story was also told in Mountstuart E. Grant Duff, *Notes from*

a Diary, 1873–188, 2 vols (London: John Murray, 1898), and repeated in *Nature* 58 (30 June 1898), 198 and in Sidney Lee, *King Edward VII: A Biography*, 2 vols (London: Macmillan, 1927), 2: 408.

14. Edward FitzGerald to E. B. Cowell, [24 July 1847], in A. M. Terhune and A. B. Terhune (eds), *The Letters of Edward FitzGerald*, 4 vols (Princeton: Princeton University Press, 1980), 1: 566.
15. John F. W. Herschel, 'An Address to the Subscribers to the Windsor and Eton Public Library and Reading Room', in *Essays from the Edinburgh and Quarterly Reviews, with Addresses and other Pieces* (London: Longman, 1857), 1–20, at p. 13.
16. 'Lyell's Principles of Geology', *Spectator* (14 January 1832), 39–40, at p. 39. For more general reflections on the revolution in manners, see Vic Gatrell, *City of Laughter: Sex and Satire in Eighteenth-Century London* (London: Atlantic Books, 2006).
17. [Thomas De Quincey], 'Pope', *North British Review* 9 (August 1848), 299–333, at pp. 301–2; and [Thomas De Quincey], 'Letters to a Young Man whose Education has been Neglected', *London Magazine* 7 (January–May 1823), 84–90, 189–94, 325–35, 556–8, esp. 332–5; 8 (July 1823), 87–95, esp. p. 87. Charlotte Sleigh, *Literature and Science* (Basingstoke: Palgrave Macmillan, 2011) provides an accessible overview of the issues involved in considering different modes of writing in relation to literary form. For further perspectives, see James J. Bono et al., 'Making Knowledge: History, Literature, and the Poetics of Science', *Isis* 101 (2010), 555–98, and the introduction and essays in Ben Marsden, Hazel Hutchinson, and Ralph O'Connor (eds), *Uncommon Contexts: Encounters between Science and Literature, 1800–1914* (London: Pickering & Chatto, 2013), 111–23.
18. [De Quincey], 'Pope', 303–4.

BIBLIOGRAPHY OF WORKS PUBLISHED AFTER 1900

Alborn, Timothy L. 'The Business of Induction: Industry and Genius in the Language of British Scientific Reform, 1820–1840', *History of Science* 34 (1996): 91–121.

Allen, David. *The Naturalist in Britain: A Social History*. Harmondsworth: Penguin, 1976.

Allen, David. *Books and Naturalists*. London: Collins, 2010.

Altick, Richard D. *The English Common Reader: A Social History of the Mass Reading Public, 1800–1900*. Chicago: University of Chicago Press, 1957.

Arianrhod, Robyn. *Seduced by Logic: Émilie Du Châtelet, Mary Somerville and the Newtonian Revolution*. Oxford: Oxford University Press, 2012.

Ashton, Rosemary. *Victorian Bloomsbury*. New Haven and London: Yale University Press, 2012.

Ashworth, William J. 'The Calculating Eye: Baily, Herschel, Babbage and the Business of Astronomy', *British Journal for the History of Science* 27 (1994): 409–41.

Baldwin, Melina Clare. *Making* Nature: *The History of a Scientific Journal*. Chicago: University of Chicago Press, forthcoming.

Baldwin, Stuart A. *Charles Lyell: A Brief Bibliography*. Wickham Bishops, Witham: Baldwin's Scientific Books, 2013.

Barnard, John, et al. (eds). *The Cambridge History of the Book in Britain*, 7 vols. Cambridge: Cambridge University Press, 1999–.

Bartholomew, Michael. 'Lyell and Evolution', *British Journal for the History of Science* 6 (1973): 261–303.

Barton, Ruth. 'John Tyndall: Pantheist: A Rereading of the Belfast Address', *Osiris*, 2nd ser., 3 (1987): 111–34.

Barton, Ruth. "Men of Science": Language, Identity, and Professionalization in the Mid-Victorian Scientific Community', *History of Science* 41 (2003), 73–119.

Beer, Gillian. *Open Fields: Science in Cultural Encounter*. Oxford: Clarendon Press, 1996.

Bennett, Scott. 'John Murray's Family Library and the Cheapening of Books in Early Nineteenth Century Britain', *Studies in Bibliography* 29 (1976): 139–66.

Berg, Maxine. *The Machinery Question and the Making of Political Economy, 1815–1848*. Cambridge: Cambridge University Press, 1980.

Bono, James J., et al. 'Making Knowledge: History, Literature, and the Poetics of Science', *Isis* 101 (2010): 555–98.

Bord, Joe. *Science and Whig Manners: Science and Political Style in Britain, c.1790–1850*. Basingstoke: Palgrave Macmillan, 2009.

Brock, Claire. 'The Public Worth of Mary Somerville', *British Journal for the History of Science* 39 (2006): 255–72.

Browne, Janet. *Charles Darwin: Voyaging*. London: Jonathan Cape, 1995.

Buckland, Adelene. *Novel Science: Fiction and the Invention of Nineteenth-Century Geology*. Chicago: University of Chicago Press, 2013.

Burkhardt, Frederick, et al. (eds). *The Correspondence of Charles Darwin*. Cambridge: Cambridge University Press, 1985–.

Burnett, John (ed.). *Destiny Obscure: Autobiographies of Childhood, Education and Family from the 1820s to the 1920s*. London: Allen Lane, 1982.

Butler, Marilyn. 'Introduction', in Mary Shelley, *Frankenstein or the Modern Prometheus: The 1818 Text*. Oxford: Oxford University Press, 1994.

Buttmann, Günther. *The Shadow of the Telescope: A Biography of John Herschel*. New York: Charles Scribner's Sons, 1970.

Cannon, Susan Faye. *Science in Culture: The Early Victorian Period*. Folkestone: Dawson, 1978.

Cannon, Walter F. 'The Impact of Uniformitarianism', *Proceedings of the American Philosophical Society* 105 (1961): 301–14.

Cannon, Walter F. 'John Herschel and the Idea of Science', *Journal of the History of Ideas* 22 (1961): 215–39.

Carlyle, Thomas. *Sartor Resartus: The Life and Opinions of Herr Teufelsdröckh in Three Books*, ed. Roger L. Tarr. Berkeley and Los Angeles: University of California Press, 2000.

Chandler, James. *England in 1819: The Politics of Literary Culture and the Case of Romantic Historicism*. Chicago: University of Chicago Press, 1998.

Chapman, Alan. *The Victorian Amateur Astronomer: Independent Astronomical Research in Britain, 1820–1920*. London: John Wiley, 1998.

Clark, Lorna J. (ed.). *The Letters of Sarah Harriet Burney*. Athens, Georgia: University of Georgia Press, 1997.

Clarke, John M. *James Hall of Albany, Geologist and Paleontologist, 1811–1898*. Albany, 1921.

Cooter, Roger. *The Cultural Meaning of Popular Science: Phrenology and the Organization of Consent in Nineteenth-Century Britain*. Cambridge: Cambridge University Press, 1984.

Cooter, Roger and Stephen Pumpfrey. 'Separate Spheres and Public Places: Reflections on the History of Science Popularization and Science in Popular Culture', *History of Science* 32 (1994): 237–67.

Copeland, Edward. *The Silver Fork Novel: Fashionable Fiction in the Age of Reform.* Cambridge: Cambridge University Press, 2012.

Corsi, Pietro. 'The Heritage of Dugald Stewart: Oxford Philosophy and the Method of Political Economy', *Nuncius* 2 (1987) 89–144.

Crowe, Michael J., et al. (eds). *A Calendar of the Correspondence of Sir John Herschel.* Cambridge: Cambridge University Press, 1998.

Cunningham, Andrew and Perry Williams. 'De-centring the "Big Picture": *The Origins of Modern Science* and the Modern Origins of Science', *British Journal for the History of Science* 26 (1993): 407–32.

Darwin, Charles. *Evolutionary Writings*, ed. James A. Secord. Oxford: Oxford University Press, 2008.

Dean, Dennis. '"Through Science to Despair": Geology and the Victorians', in James Paradis and Thomas Postlewait (eds), *Victorian Science and Victorian Values.* New York: New York Academy of Sciences, 1985: 111–36.

Dean, Dennis. *James Hutton and the History of Geology*. Ithaca, New York: Cornell University Press, 1992.

Desmond, Adrian. 'The Making of Institutional Zoology in London, 1822–1836', *History of Science* 23 (1985): 153–85, 223–50.

Desmond, Adrian. *The Politics of Evolution: Morphology, Medicine, and Reform in Radical London.* Chicago: University of Chicago Press, 1989.

Donald, Diana. *The Age of Caricature: Satirical Prints in the Reign of George III.* New Haven: Yale University Press, 1996.

Donald, Diana. *Followers of Fashion: Graphic Satires from the Georgian Period.* London: Hayward Gallery, 2002.

Fielding, K. J. and Rodger L. Tarr (eds). *Carlyle Past and Present: A Collection of New Essays.* London: Vision Press, 1976.

Flaherty, Clare. 'A Recently Rediscovered Unpublished Manuscript: The Influence of Sir Humphry Davy on Anne Brontë', *Brontë Studies* 38 (2013): 30–41.

Fullmer, June Z. *Sir Humphry Davy's Published Works.* Cambridge, Massachusetts: Harvard University Press, 1969.

Fullmer, June Z. *Young Humphry Davy: The Making of an Experimental Chemist.* Philadelphia: American Philosophical Society, 2000.

Fyfe, Aileen. *Science and Salvation: Evangelicals and Popular Science Publishing in Victorian Britain.* Chicago: University of Chicago Press, 2004.

Fyfe, Aileen. *Steam-Powered Knowledge: William Chambers and the Business of Publishing, 1820–1860.* Chicago: University of Chicago Press, 2012.

Fyfe, Aileen and Bernard Lightman (eds). *Science in the Marketplace: Nineteenth-Century Sites and Experiences.* Chicago: University of Chicago Press, 2007.

Gatrell, Vic. *City of Laughter: Sex and Satire in Eighteenth-Century London.* London: Atlantic Books, 2006.

'George Acorn' (pseudonym). *One of the Multitude.* London: W. Heinemann, 1911.

Gleason, Mary Louise. *The Royal Society of London, 1827–1847*. New York: Garland Publishing, 1991.

Golinski, Jan. *Science as Public Culture: Chemistry and Enlightenment in Britain, 1760–1820*. Cambridge: Cambridge University Press, 1992.

Golinski, Jan. 'Humphry Davy's Sexual Chemistry', *Configurations* 7 (1999): 15–41.

Golinski, Jan. 'Humphry Davy: The Experimental Self', *Eighteenth-Century Studies* 43 (2011): 15–28.

Gould, Stephen Jay. *Time's Arrow, Time's Cycle*. Cambridge, Massachusetts: Harvard University Press, 1987.

Gray, Valerie. *Charles Knight: Educator, Publisher, Writer*. Aldershot, 2006.

Hall, Marie Boas. *All Scientists Now: The Royal Society in the Nineteenth Century*. Cambridge: Cambridge University Press, 1984.

Hamilton, James. *London Lights: The Minds that Moved the City that Shook the World, 1805–51*. London: John Murray, 2007.

Harrison, J. F. C. *The Second Coming: Popular Millenarianism, 1780–1850*. London: Routledge, 1979.

Hay, William. *The Whig Revival, 1808–1830*. Basingstoke: Palgrave Macmillan, 2004.

Hays, J. N. 'The Rise and Fall of Dionysius Lardner', *Annals of Science* 38 (1981): 527–42.

Herbert, Sandra. *Charles Darwin, Geologist*. Ithaca, New York: Cornell University Press, 2005.

Heringman, Noah. *Romantic Rocks, Aesthetic Geology*. Ithaca, New York: Cornell University Press, 2004.

Hill, Alan G. and Ernest de Selincourt (eds). *The Letters of William and Dorothy Wordsworth*, 2nd ed. Oxford: Clarendon Press, 1979.

Hilton, Boyd. *The Age of Atonement: The Influence of Evangelicalism on Social and Economic Thought, 1795–1865*. Oxford: Clarendon Press, 1988.

Hilton, Boyd. *A Mad, Bad, and Dangerous People? England, 1783–1846*. Oxford: Oxford University Press, 2006.

Hollis, Patricia. *Class and Conflict in Nineteenth Century England: 1815–1850*. London: Routledge, 1973.

Holmes, Richard. *The Age of Wonder: How the Romantic Generation Discovered the Beauty and Terror of Science*. London: HarperCollins, 2008.

Horsley, Victor. 'Description of the Brain of Mr. Charles Babbage, F.R.S.', *Philosophical Transactions of the Royal Society of London* 200 (1909): 117–31.

Hyman, Anthony. *Charles Babbage: Pioneer of the Computer*. Oxford: Oxford University Press, 1984.

Jenkins, Alice. *Space and the 'March of Mind': Literature and the Physical Sciences in Britain, 1815–1850*. Oxford: Oxford University Press, 2007.

Jenkins, Alice. 'Genre and Geometry: Victorian Mathematics and the Study of Literature and Science', in Ben Marsden, Hazel Hutchinson, and Ralph O'Connor (eds), *Uncommon Contexts: Encounters between Science and Literature, 1800-1914*. London: Pickering & Chatto, 2013: 111–23.

Johns, Adrian. 'The Identity Engine: Printing and Publishing at the Beginning of the Knowledge Economy', in Lissa Roberts, Simon Schaffer, and Peter Dear (eds), *The Mindful Hand: Inquiry and Invention from the Late Renaissance to Early Industrialisation*. Amsterdam: Koninklijke Nederlandse Akademie van Wetenschappen, 2007: 403–28.

Klancher, Jon P. *The Making of English Reading Audiences, 1790–1832*. Madison: University of Wisconsin Press, 1987.

Knight, David. *Humphry Davy: Science and Power*. Cambridge: Cambridge University Press, 1996.

Koselleck, Reinhart. *Futures Past: On the Semantics of Historical Time*. New York: Columbia University Press, 2004.

Lamb, John B. '"Spiritual Enfranchisement": *Sartor Resartus* and the Politics of *Bildung*', *Studies in Philology* 107 (2010): 259–82.

Latané, David E. *William Maginn and the British Press: A Critical Biography*. Farnham: Ashgate, 2013.

Laudan, Rachel. 'The Role of Methodology in Lyell's Science', *Studies in History and Philosophy of Science* 13 (1982): 215–49.

Laudan, Rachel. *From Mineralogy to Geology: The Foundations of a Science, 1650–1830*. Chicago: University of Chicago Press, 1987.

Leask, Nigel. *British Romantic Writers and the East: Anxieties of Empire*. Cambridge: Cambridge University Press, 1992.

Lee, Sidney. *King Edward VII: A Biography*, 2 vols. London: Macmillan, 1927.

Levine, George and William Madden (eds), *The Art of Victorian Prose*. New York: Oxford University Press, 1968.

Lightman, Bernard. *Victorian Popularizers of Science: Designing Nature for New Audiences*. Chicago: University of Chicago Press, 2007.

Lyell, Charles. *Principles of Geology*, ed. James A. Secord. London: Penguin Books, 1997.

MacCalman, Iain. *Radical Underworld: Prophets, Revolutionaries and Pornographers in London, 1795–1840*. Cambridge: Cambridge University Press, 1988.

MacMillan, Dorothy (ed.). *Queen of Science: Personal Recollections of Mary Somerville*. Edinburgh: Canongate Classics, 2001.

McPhee, John. *Basin and Range*. New York: Farrar, Straus and Giroux, 1981.

Mandler, Peter. *Aristocratic Government in the Age of Reform: Whigs and Liberals, 1830–1852*. Oxford: Clarendon Press, 1990.

Marsden, Ben, Hazel Hutchinson, and Ralph O'Connor (eds). *Uncommon Contexts: Encounters between Science and Literature, 1800–1914*. London: Pickering & Chatto, 2013.

Maxwell, Richard and Katie Trumpener (eds). *The Cambridge Companion to Fiction in the Romantic Period*. Cambridge: Cambridge University Press, 2008.

Mellor, Anne K. *English Romantic Irony*. Cambridge, Massachusetts: Harvard University Press, 1980.

Miller, David Philip. 'Between Hostile Camps: Sir Humphry Davy's Presidency of the Royal Society of London, 1820–1827', *British Journal for the History of Science* 16 (1983): 1–43.

Moore, Carlisle. 'Carlyle: Mathematics and "Mathesis"', in K. J. Fielding and Rodger L. Tarr (eds). *Carlyle Past and Present: A Collection of New Essays*. London: Vision Press, 1976: 61–95.

Morrell, Jack and Arnold Thackray. *Gentlemen of Science: Early Years of the British Association for the Advancement of Science*. Oxford: Clarendon Press, 1981.

Morrell, Jack and Arnold Thackray (eds), *Gentlemen of Science: Early Correspondence of the British Association for the Advancement of Science*. London: Royal Historical Society, 1984.

Morus, Iwan Rhys. 'Correlation and Control: William Robert Grove and the Construction of a New Philosophy of Scientific Reform', *Studies in the History and Philosophy of Science* 22 (1991): 589–621.

Morus, Iwan Rhys. *Frankenstein's Children: Electricity, Exhibition, and Experiment in Early Nineteenth-Century London*. Princeton: Princeton University Press, 1993.

Morus, Iwan Rhys. *When Physics became King*. Chicago: University of Chicago Press, 2005.

Neeley, Kathryn A. *Mary Somerville: Science, Illumination, and the Female Mind*. Cambridge: Cambridge University Press, 2001.

Oberg, Barbara B. et al. (eds). *The Papers of Thomas Jefferson*, vol. 32. Princeton: Princeton University Press, 2005.

O'Connor, Ralph. *The Earth on Show: Fossils and the Poetics of Popular Science, 1802–1856*. Chicago: University of Chicago Press, 2007.

O'Connor, Ralph. 'Young-Earth Creationists in Early Nineteenth-Century Britain? Towards a Reassessment of "Scriptural Geology"', *History of Science* 45 (2007): 357–403.

O'Connor, Ralph (ed.). *Victorian Science and Literature*, vol. 7: *Science as Romance*. London: Pickering and Chatto, 2012.

Ospovat, Dov. 'Lyell's Theory of Climate', *Journal of the History of Biology* 10 (1977): 317–39.

Parker, Mark. *Literary Magazines and British Romanticism*. Cambridge: Cambridge University Press, 2000.

Parry, Jonathan. *The Rise and Fall of Liberal Government in Victorian Britain*. New Haven and London: Yale University Press, 1996.

Patterson, Elizabeth Chambers. *Mary Somerville and the Cultivation of Science, 1815–1840*. Boston: Martinus Nijhoff, 1983.

Pearce, Edward. *Reform! The Fight for the 1832 Reform Act*. London: Pimlico, 2004.

Pearl, Sharrona. *About Faces: Physiognomy in Nineteenth-Century Britain*. Cambridge, Massachusetts: Harvard University Press, 2010.

Pickstone, John. 'Museological Science? The Place of the Analytical/Comparative in Nineteenth-Century Science, Technology, and Medicine', *History of Science* 32 (1994): 111–38.

Porter, Roy. *The Making of Geology: Earth Science in Britain, 1660–1815*. Cambridge: Cambridge University Press, 1977.

Porter, Roy. 'Charles Lyell: The Public and Private Faces of Science', *Janus* 69 (1982): 29–50.

Poston, Lawrence. 'Millites and Millenarians: The Context of Carlyle's "Signs of the Times"', *Victorian Studies* 26 (1983): 381–406.

Price, Leah. *The Anthology and the Rise of the Novel: From Richardson to George Eliot*. Cambridge: Cambridge University Press, 2000.

Rauch, Alan, *Useful Knowledge: The Victorians, Morality, and the March of Intellect*. Durham, NC, 2001.

Raven, James. *The Business of Books: Booksellers and the English Book Trade 1450–1850*. New Haven: Yale University Press, 2007.

Reingold, Nathan. 'Babbage and Moll on the State of Science in Great Britain: A Note on a Document', *British Journal for the History of Science* 4 (1968): 58–64.

Richardson, Sarah. 'Women, Philanthropy, and Imperialism in Early Nineteenth-Century Britain', in Helen Gilbert and Chris Tiffin (eds), *Burden or Benefit? Imperial Benevolence and its Legacies*. Bloomington and Indianapolis: Indiana University Press, 2008: 90–102.

Richardson, Sarah. *The Political Worlds of Women: Gender and Politics in Nineteenth Century Britain*. New York: Routledge, 2013.

Rooney, David and James Nye. '"Greenwich Observatory Time for the Public Benefit": Standard Time and Victorian Networks of Regulation', *British Journal for the History of Science* 42 (2009): 5–30.

Rose, Jonathan. *The Intellectual Life of the British Working Classes*. New Haven: Yale University Press, 2001.

Ross, Sydney. '*Scientist*: The Story of a Word', *Annals of Science* 18 (1962): 65–85.

Rudwick, Martin J. S. 'Caricature as a Source for the History of Science: De la Beche's Anti-Lyellian Sketches of 1831', *Isis* 66 (1975): 534–60.

Rudwick, Martin J. S. 'Charles Lyell, F.R.S. (1797–1875) and his London Lectures on Geology, 1832–33', *Notes and Records of the Royal Society of London* 29 (1975): 231–63.

Rudwick, Martin J. S. 'Bibliography of Lyell's Sources', in Charles Lyell, *Principles of Geology*, 3 vols. Chicago: University of Chicago Press, 1990–1: 3: 113–60.

Rudwick, Martin J. S. *Worlds Before Adam: The Reconstruction of Geohistory in the Age of Reform*. Chicago: University of Chicago Press, 2008.

Rupke, Nicolaas. *The Great Chain of History: William Buckland and the English School of Geology*. Oxford: Clarendon, 1983.

Ruse, Michael. 'Darwin's Debt to Philosophy: An Examination of the Influence of the Philosophical Ideas of John F. W. Herschel and William Whewell on the Development of Charles Darwin's Theory of Evolution', *Studies in the History and Philosophy of Science* 6 (1975): 159–81.

Russell, David. '"Our Debt to Lamb": The Romantic Essay and the Emergence of Tact', *ELH* 79 (2012): 179–209.

Sanders, C. R. and K. J. Fielding et al. (eds). *The Collected Letters of Thomas and Jane Welsh Carlyle* 1–16. Durham, North Carolina: Duke University Press, 1970–90.

Schaffer, Simon. 'Scientific Discoveries and the End of Natural Philosophy', *Social Studies of Science* 16 (1986): 387–420.

Schaffer, Simon. 'Babbage's Intelligence: Calculating Engines and the Factory System', *Critical Inquiry* 21 (1994): 203–27.

Schaffer, Simon. 'Babbage's Dancer and the Impresarios of Mechanism', in Francis Spufford and Jenny Uglow (eds), *Cultural Babbage: Time, Technology and Invention*. London: Faber, 1996: 52–80.

Schiebinger, Londa. *Nature's Body: Women in the Origins of Modern Science*. Boston: Beacon Press, 1989.

Scruton, Roger. 'Hope's Philosophical Excursus', in David Watkin and Philip Hewat-Jaboor (eds), *Thomas Hope: Regency Designer*. New Haven and London: Yale University Press, 2008: 243–7.

Secord, Anne. 'Corresponding Interests: Artisans and Gentlemen in Nineteenth-Century Natural History', *British Journal for the History of Science* 27 (1994): 383–408.

Secord, Anne. 'Science in the Pub: Artisan Botanists in Early Nineteenth-Century Lancashire', *History of Science* 32 (1994): 383–408.

Secord, Anne. 'Botany on a Plate: Pleasure and the Power of Pictures in Promoting Early Nineteenth-Century Scientific Knowledge', *Isis* 93 (2002): 28–57.

Secord, Anne. 'Introduction', in Gilbert White, *The Natural History of Selborne*. Oxford: Oxford University Press, 2013.

Secord, James A. 'The Discovery of a Vocation: Darwin's Early Geology', *British Journal for the History of Science* 24 (1991): 133–57.

Secord, James A. 'Edinburgh Lamarckians: Robert Jameson and Robert E. Grant', *Journal of the History of Biology* 24 (1991): 1–18.

Secord, James A. *Victorian Sensation: The Extraordinary Publication, Reception and Secret Authorship of* Vestiges of the Natural History of Creation. Chicago: University of Chicago Press, 2000.

Secord, James A. 'Knowledge in Transit', *Isis* 95 (2004): 654–72.

Secord, James A. 'Scrapbook Science: Composite Caricatures in Late Georgian London', in Ann B. Shteir and Bernard Lightman (eds), *Figuring it Out: Science, Gender, and Visual Culture*. Hanover, New Hampshire: Dartmouth College Press, 2006: 164–91.

Secord, James A. 'How Scientific Conversation became Shop Talk', in Aileen Fyfe and Bernard Lightman (eds), *Science in the Marketplace: Nineteenth-Century Sites and Experiences*. Chicago: University of Chicago Press, 2007: 23–59.

Secord, James A. 'Science, Technology and Mathematics', in David McKitterick (ed.), *The Cambridge History of the Book in Britain, Volume VI, 1830–1914*. Cambridge: Cambridge University Press, 2009: 443–74.

Sher, Richard B. *The Enlightenment and the Book: Scottish Authors and their Publishers in Eighteenth-Century Britain, Ireland, and America*. Chicago: University of Chicago Press, 2006.

Siskin, Clifford. *The Work of Writing: Literature and Social Change in Britain, 1700–1830*. Baltimore: Johns Hopkins University Press, 1998.

Slater Michael (ed.). *Dickens's Journalism*, 4 vols. London: J. M. Dent, 1996.

Sleigh, Charlotte. *Literature and Science*. Basingstoke: Palgrave Macmillan, 2011.

Snyder, Laura J. *Reforming Philosophy: A Victorian Debate on Science and Society*. Chicago: University of Chicago Press, 2006.

Snyder, Laura J. *The Philosophical Breakfast Club: Four Remarkable Friends who Transformed Science and Changed the World*. New York: Broadway Books, 2011.

Somerville, Mary. *Collected Works of Mary Somerville*, ed. James A. Secord, 9 vols. Bristol: Thoemmes Press, 2004.

St Clair, William. *The Reading Nation in the Romantic Period*. Cambridge: Cambridge University Press, 2004.

Stack, David. *Queen Victoria's Skull*. London: Hambleton Continuum, 2008.

Stapleford, Brian. 'Science Fiction before the Genre', in Edward James and Farah Mendlesohn (eds), *The Cambridge Companion to Science Fiction*. Cambridge: Cambridge University Press, 2003: 15–31.

Stoddart, D. R. 'Darwin, Lyell, and the Geological Significance of Coral Reefs', *British Journal for the History of Science* 9 (1976): 199–218.

Stott, Rebecca. 'Thomas Carlyle and the Crowd: Revolution, Geology and the Convulsive "Nature" of Time', *Journal of Victorian Culture* 4 (1999): 1–24.

Sumner, James. *Brewing Science, Technology and Print, 1700–1880*. London: Pickering & Chatto, 2013.

Tarr, Rodger L. 'Introduction', in Thomas Carlyle, *Sartor Resartus: The Life and Opinions of Herr Teufelsdröckh in Three Books*. Berkeley and Los Angeles: University of California Press, 2000: xxi–xciv.

Tee, Gary J. 'Relics of Davy and Faraday in New Zealand', *Notes and Records of the Royal Society of London* 52 (1998): 93–102.

Tennyson, G. B. *Sartor called Resartus: The Genesis, Structure, and Style of Thomas Carlyle's First Major Work*. Princeton: Princeton University Press, 1965.

Terhune, A. M. and A. B. Terhune (eds). *The Letters of Edward FitzGerald*, 4 vols. Princeton: Princeton University Press, 1980.

Thompson, E. P. *The Making of the English Working Class*. London: Penguin Books, 1980.

Thrall, Miriam M. H. *Rebellious Fraser's: Nol Yorke's Magazine in the Days of Maginn, Thackeray, and Carlyle*. New York: Columbia University Press, 1934.

Topham, Jonathan. 'Science and Popular Education in the 1830s: The Role of the *Bridgewater Treatises*', *British Journal of the History of Science* 25 (1992): 397–430.

Topham, Jonathan. 'Beyond the "Common Context": The Production and Reading of the *Bridgewater Treatises*', *Isis* 89 (1998): 233–62.

Topham, Jonathan. 'Scientific Publishing and the Reading of Science in Nineteenth-Century Britain: A Historiographical Survey and Guide to Sources', *Studies in the History and Philosophy of Science* 31 (2000): 559–612.

Topham, Jonathan. 'John Limbird, Thomas Byerley, and the Production of Cheap Periodicals in the 1820s', *Book History* 8 (2005): 75–106.

Topham, Jonathan, et al. 'Historicizing "Popular Science"', *Isis* 100 (2009): 310–68.

Topham, Jonathan. *Reading the Book of Nature: Science, Religion and the Culture of Print in the Age of Reform*. Forthcoming.

Trela, D. J. and Rodger L. Tarr (eds). *The Critical Response to Thomas Carlyle's Major Works*. Westport, Connecticut: Greenwood Press, 1997.

Tresch, John. *The Romantic Machine: Utopian Science and Technology after Napoleon*. Chicago: University of Chicago Press, 2012.

Turner, Frank M. *Contesting Cultural Authority: Essays in Victorian Intellectual Life*. Cambridge: Cambridge University Press, 1993.

Ulrich, John M. 'Thomas Carlyle, Richard Owen, and the Paleontological Articulation of the Past', *Journal of Victorian Culture* 11 (2006): 30–56.

Van Wyhe, John. *Phrenology and the Origins of Victorian Scientific Naturalism*. Aldershot: Ashgate, 2004.

Vincent, David. *Literacy and Popular Culture*. Cambridge: Cambridge University Press, 1989.

Wallbank, Adrian J. *Dialogue, Didacticism and the Genres of Dispute: The Literary Dialogue in an Age of Revolution*. London: Pickering & Chatto, 2012.

Walters, Alice. 'Conversation Pieces: Science and Politeness in Eighteenth-Century England', *History of Science* 35 (1997): 121–54.

Waring, Sophie. 'Thomas Young and the Board of Longitude', abstract of paper delivered at British Society for the History of Science annual meeting, 2011.

Webb, R. K. *The British Working Class Reader, 1790–1848: Literacy and Social Tension*. London: George Allen & Unwin, 1955.

White, Paul. *Thomas Huxley: Making the 'Man of Science'*. Cambridge: Cambridge University Press, 2003.

Wilson, Leonard G. (ed.). *Sir Charles Lyell's Scientific Journals on the Species Question.* New Haven and London: Yale University Press, 1970.

Wilson, Leonard G. *Charles Lyell: The Years to 1841: The Revolution in Geology.* New Haven and London: Yale University Press, 1972.

Winter, Alison. *Mesmerized: Powers of Mind in Victorian Britain.* Chicago: University of Chicago Press, 1998.

Wood, Alexander. *Thomas Young: Natural Philosopher, 1773–1829.* Cambridge: Cambridge University Press, 1954.

Wordsworth, William. *The Major Works,* ed. Stephen Gill. Oxford: Oxford University Press, 1984.

Yeo, Richard. *Defining Science: William Whewell, Natural Knowledge, and Public Debate in Early Victorian Britain.* Cambridge: Cambridge University Press, 1993.

INDEX